VIDEO PROJECTION 101

The Pre-Production and Execution
STRATEGIES
of a Video Projectionist

Clem **Harrod**

CLEMCO.U
Tampa, FL

Copyright © 2020 by Clem Harrod

All rights reserved. No part of this publication may be reproduced, distributed, or transmitted in any form or by any means, including photocopying, recording, or other electronic or mechanical methods, without the prior written permission of the publisher, except in the case of brief quotations embodied in critical reviews and certain other noncommercial uses permitted by copyright law. For permission requests, write to the publisher, addressed "Attention: Permissions Coordinator," via this website: www.clemcou.com | @clemcoav

Ordering Information:

Quantity sales. Special discounts are available on quantity purchases by schools, corporations, associations, and others. For details, contact the publisher at www.clemcou.com. Orders by U.S. trade bookstores and wholesalers. Please contact: CLEMCO.U at www.clemcou.com.

Printed in the United States of America

This publication is designed to provide competent and reliable information regarding the subject matter covered. However, it is sold with the understanding that the author and publisher are not engaged in rendering legal, financial, or other professional advice. Laws and practices often vary from state to state and if legal or other expert assistance is required, the services of a professional should be sought. The author and publisher specifically disclaim any liability incurred from the use or application of the contents of this book.

Events, locales, and conversations have been recreated from the author's best memories of them. In some instances, to maintain the anonymity of individuals mentioned in this publication, the author has changed names, places, identifying characteristics and details such as physical properties, occupations, and places of residence.

First Printing: 2022

ISBN: 978-1-7347452-2-1

Cover design by CCS Marketing

Cover ballroom layout designed by Tom Bollard

Edited by Shirley Jump

Book design by Christina Minopoli

Back Cover photo taken by Donovan Muir

"To the youths who march onward and upward toward the light,
this volume is respectively dedicated."
—Charles H. Wesley

"For generational knowledge and understanding is one of the most powerful
things in the world. However, the lack thereof is one of the deadliest."
—Clem Harrod

To every mentor I've had, to every person I mentored, this book is for you. I would not be who I am without your support, and I wouldn't be where I am without your willingness to share your insight.

A piece of art may be beautiful on its own, but when curated with other bodies of work, a gallery is born.

Thank you!

Acknowledgements

To my loving and supportive wife, Joslynne. Thank you for your patience as I grew in the understanding of self and my God-given gift. This journey has not been easy, and knowing that you, CJ, and Kinley were always there cheering for me and wanting me to come home is what makes this never-ending story possible. I see it…

Mom, sight beyond sight, faith beyond faith, and wisdom beyond wisdom. Those are the things you have shown me, and things I pray I am able to show others. I believe it…

To a few of the many Video Projectionists, Video Engineers, Lighting Designers, Master Electricians, Audio Engineers, Project Managers, Technical Directors, Production Companies, Rental & Staging Companies, and Industry Leaders who have impacted my career. Thank you for sharing your skills, techniques, beliefs, jobs, opportunities, wisdom, and overall love with me. WE'VE achieved it!

Phil Licari, Melvin LeGrand, Mike Swinton, Stuart Brown, Zamir Zeigler, Steve Campbell, Steve Olson, Danny Harris, Jerry Farnum, Harry Wilkins, Kevin Rose, Drew Brome, Robert Permenter, Ben Standfield, Matt Ellis, Rick Wegner, Rudy Tessmer, Bob Satmary, Ken Wisniski, Chris Rotkosky, Tim Durr, James Leslie, Sean Borowski, Kevin McCabe, Jack McCabe, Steve Micio, Darrin Peterson, Brian Lingo, Jason Lettow, John Brewer, Richard Dunn, Andrew Douglas, Heather Crowne, Eric Durning, Jay Merriam, Billy Walsh, Jonathan Clark, Jay Richardson, Beth Forbes, Tom Bollard, Mike Anderson, Keith Elliott, Tom Kervitsky, Mike Compton, Kari Hyatt, Nick Farrell, Scott Flanagan, Leizl Bala, Tammy Scala, TJ Kervitsky, Les Goldberg, Neil Morrison, Craig Muir, Erika Dingman, Amadel Smith, Joe Freeman, Ricardo Hicks, Tyler Mayne, Bob Murdock, Omar Colom, Carlton "CJ" Moment, Ethan Holder, Rick Price, Steve Uhlmer.

To an INCREDIBLE team! CLEMCO.U was able to publish its second book because of you. There was no way I was going to carry this ball across the finish line without you ALL pushing, pulling and encouraging me to get it done. Thank you for your love and support.

Shirley Jump, Dawn Bedingfield, Christina Minopoli, Jason Crawford, Crystal Crawford, Sharri Hall, Bridgette Thomas, Bruce Couch, Donovan Muir, Rich Johnson, Barbara Henderson, Willard Henderson, Jillian Session, Marcus Session, Jamal Brown, Georgie Delgado, Megan Morris, Lesroy Louard, Ashley Lamb, Christian Cisar, Chris "Chiili" Horton, Ben Poe, Chris Poe, Toriano Evans, Nickita Harrison, and everyone who placed a pre-order of VP101.

#Projection101

Foreword by Nick Farrell

Persistence of Vision is the phenomenon that allows the human eye to perceive movement in motion pictures. We display individual images at 30 frames per second and our mind sees fluid movement. Persistence of Vision is also an expression that I would use to describe some of the best attributes of Clem Harrod.

As an Executive Producer and Creative Director for corporate production, I have had the opportunity to work closely with Clem over the last twenty years. His approach to his work embodies both Persistence and Vision.

Clem brings a curiosity in his approach to every project that separates him from many others. I can always expect questions from technicians representing the different disciplines represented on a project, but oftentimes Clem's questions run deeper. It's not only questions about the number of screens, or the positioning of screens in the room, it's questions that strike at the very reason we are gathering people in a room together to begin with. "What is the messaging goal of this meeting?" "Is this a celebratory event or an educational learning experience?" In response, one might say "what does that have to do with putting an image on a screen?" That is where Clem's Persistent curiosity allows him to understand the greater Vision for the event.

It was clear to me from early on that Clem had a higher calling than the pursuit of a career as an event technician. That career alone is admirable, profitable, and fulfilling on many levels. The technicians that work tirelessly (oftentimes unrecognized or celebrated) are the heroes of the industry, working unseen to the thousands of people whose lives they have touched. But Clem saw a greater purpose in the work he was doing. He saw how his life and the lives of many of his colleagues had benefited from being a part of this "hidden industry". He decided that he would make sure that others could have the same opportunity. First and foremost though he wanted people to be armed with the knowledge that would help them succeed.

Clem realized early on that his professional and life experiences put him in a unique position to help others through mentoring and training. Many have similar experiences to pull from, but not many have the patience, drive, and compassion it takes to make elevating others their mission.

This became clear to me in one of our many "post-workday" conversations. I remember sitting in a hotel lobby alcove in Carlsbad CA when it became clear to me that Clem had higher aspirations than just being the best projectionist he could be. He wanted to be the best person he could be.

Career Projection 101 was Clem's first book and one that I wish I had available to me when I set out as a freelancer many years ago. *Video Projection 101* will prove to be another invaluable resource to anyone pursuing a career in the Production industry. The themes contained are a combination of building blocks for a better career, and also an instruction manual on how to succeed in work and in life.

Video Projection 101 is a very personal and unique instructional guide shared by a person who has walked the walk. The lessons are applicable across industries. One specific theme that stands out for me is understanding your place as part of a bigger picture. Clem wants readers to understand clearly that they are unique and integral to the success of any project. Clem references Nelson Mandela's philosophy of Ubuntu as a way of describing a person's role in a project. If you are invited to work on an a project you have a unique opportunity bestowed upon you to bring your individual talents and spirit to that project as part of a greater team. Working side by side with others, asking questions, sharing information, having each other's backs makes us collectively better at what we do and ultimately delivers the most successful projects.

Another takeaway from the book is the importance of relationships. At one point in the book Clem refers to it as having "history". Relationships with a client or with an event are the key to a successful career in the industry. The deeper the relationship the better you understand what is expected of you. Each event you work on is an opportunity to start a relationship that may result in you working for that same client year after year. Repeat business is the ultimate compliment.

I have no doubt that this book will become a critical resource to those who want to live better, work better, and experience an enriching life of curiosity and fulfillment in Production or any other professional pursuit they undertake.

—Nick Farrell, Vice President/Executive Producer, TEK Productions

Foreword by Omar Colom

I met Clem Harrod in 2015, after working in the Rental & Staging Industry for several years. I originally started in the carpentry department and worked my way up to a Lead Video Engineer. That wasn't an easy task in those days because information about the Audio-Visual Industry wasn't readily available.

However, Clem's mentorship changed everything. I've had some great mentors in this industry, but Clem was different. I had been looking for a way to network with other people in the industry while also growing my skills and expanding my expertise when I came across the Video Projection Support Group (VPSG) on Facebook.

This tiny yet powerful group had thousands of members who willingly and graciously shared their advice and experience whenever anyone asked a question. The best part for me, however, was the Video Projection 101 Practicum Class that Clem offered. I was able to meet Clem in person, and I was very impressed.

When I came to the class, I was already a skilled Video Projectionist but knew there was always something new to be learned. I liked seeing how someone else worked out the problems we all run into on the job. Watching Clem explain and show his workflow was more like watching art being created rather than the sharing of technical information. He brought color and dimension to something that had always been black and white for me.

I had come into class wanting to know the Standard Operating Procedures for setting up a project. All my mentors had given me great information—but much of it was also contradictory. I wanted a definitive answer. Clem made it all so simple, helping me see that there is no one right way for anything because every projector, venue, screen, and most importantly, every team, is different and each one requires a different approach.

More than that, Clem gave me the confidence to believe in my skills. When I approached him again several weeks later, looking for something I needed but couldn't find, Clem's answer was simple: "Why don't you create the resource you're looking for?" Shortly after that, I started AV Educate as a resource for helping other freelancers learn about upcoming classes and training opportunities.

I feel fortunate to be able to call Clem a friend, both professionally and personally. If you follow his path, you will see the building blocks to success. His teaching manner balances storytelling, knowledge and easy-to-understand technical information. Even if you are an experienced veteran in the industry, there are stories you can relate to in this book and information you can learn. I have read dozens of books about the Audio-Visual Industry, and none of them are even close to the caliber of Clem's books.

Woven into the fabric of every word is Clem's clear commitment to the community that works with and around him, and his belief in the Ubuntu philosophy—I am because we are. Clem believes in and encourages everyone in the industry to rise to a new level of professionalism while also maintaining camaraderie and promoting everyone to be their best.

Through *VP101*, you will learn the hard skills needed to be a Video Projectionist but also, and just as critical, the soft skills you need to succeed as a Video Projectionist. It is clear to me that Clem is committed to improving the AV Industry, and anyone who reads this book will be a part of that commitment too.

At the end of the day, as Billy Joel once said, "You're only as good as your crew."

—Omar Colom, Director of Education, Evolve Technology; Founder, AV Educate

Interview

On November 15, 2021 Dawn Bedingfield sat down to chat with the author, Clem Harrod:

Dawn: Could you name one of the greatest challenges that you've experienced in video production, either in general or in your career, and one of your favorite memories?

Clem: One of my biggest challenges was having the confidence in myself. To see the value of what I bring [to the table] through [my] ability to understand and see shapes and lines. And to have that dedication to wanting to make those shapes and lines all align.

[I knew that] I had to create a process for sustainability. I was working in a right brain artistic sense in doing the work but I understood that a process had to be established in order to allow myself to more effectively guide others to do the work that we needed to do to accomplish the job.

It wasn't easy to sit down and think about what I do, how I do it, why I do it, and then present that in a way that was digestible for someone else to learn from and to be successful at.

That was a huge challenge with a lot of introspective deep thought. It was a time-consuming process to go deep within to find these answers, and then bring them to the surface to help myself and to help others. A huge reward from that was being able to provide this structure, these tips, and these techniques in a way that I've been able to see someone progress through their career.

One of the contractors that I work with on a regular basis uses our CLEMCO platform to help find jobs and for support in managing his billing infrastructure as an independent contractor. He first started by paying me and traveling down from the Washington, DC market to Tampa, Florida. He paid his way and paid me to guide him through the book that I wrote (that is now my workbook), and to teach him what I know.

And [for him to now be] able to use these tools as a resource to change the trajectory of his career, and to increase his earning potential, and to be able to support himself and his family more comfortably, even bypassing all my knowledge—that's one of the greatest rewards.

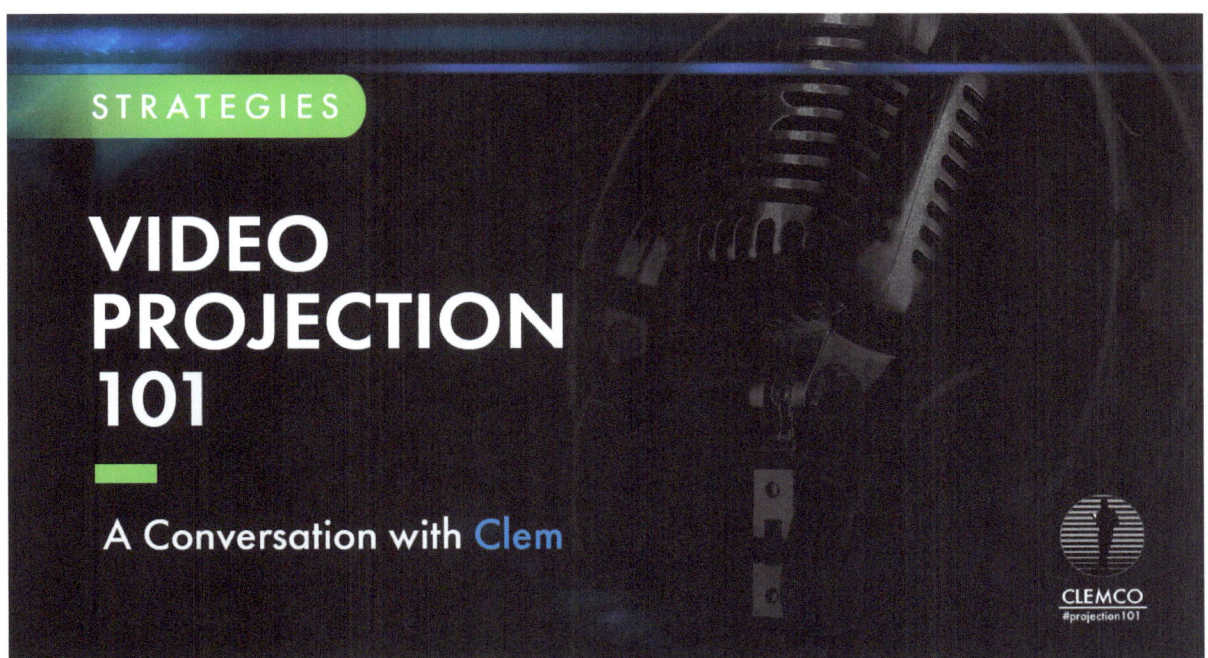

 For more of this interview and more behind-the-scenes information about the book, scan the QR Code!

Introduction

Video Projection wasn't a trade I knew existed when I studied Television Production in middle school, high school, or college. In fact, it wasn't until I was twenty-one that I realized there was a sector of my industry that created and produced live events like concerts, conventions, exhibits, and festivals. However, when I was introduced to that part of the Live Event Production Industry, it sparked my interest and it was something I wanted to pursue.

In February of 2002, I didn't know what specialty I would eventually choose or who I would eventually become, but I embraced the idea of building something out of nothing. I embraced the ability to understand one part and its function, then the ability to add another part to make it stronger and better. The combination of those two parts make a whole, and they work together to create a reaction. The Corporate Events sector of the Live Events Production Industry created a reaction within me, and I discovered engineering.

According to the Oxford Dictionary, "engineering is the branch of science and technology concerned with the design, building, and use of engines, machines, and structures…working artfully to bring something about." By definition, the way we, as individuals, come together, as a team, to build structures and experiences inside of empty shells with cases and cases of equipment and miles of cables, is engineering in its truest form. We all play a role in this, and the one I chose to express my art was that of a Video Projectionist.

As I grew in my craft, I wanted to execute my jobs and lead my teams as effectively and efficiently as possible. In order to do so, I needed to come up with a Standard Operating Procedure or SOP. I felt that was the only way to streamline my process and take what was once muscle memory and turn it into a teachable and scalable process.

To make this vision a reality, I needed to go deep within my mind and my soul to understand what I do, and why I do it. I needed to see how the gears and cogs turned in my head and what was the driving force behind it all. I needed to discover what made me me. That was the only way I was going to see myself and understand what made my procedure desirable by production companies, Technical Directors, and End Clients throughout the industry.

I needed to go on a journey to discover Projection 101, and this is what I found.

The Production Channel is a blog, podcast, and resource for those who work in the live events production industry. To learn more, scan the QR code!

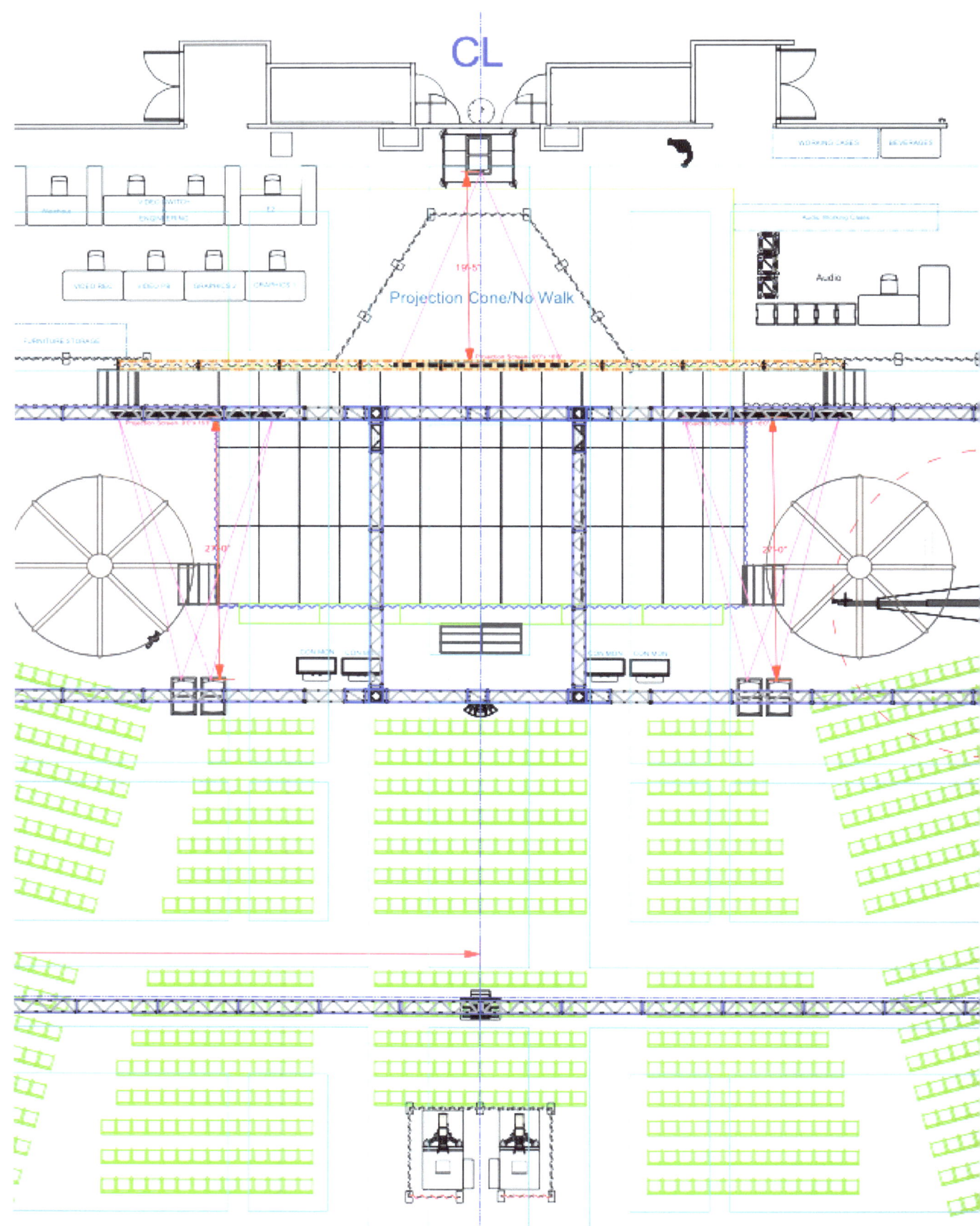

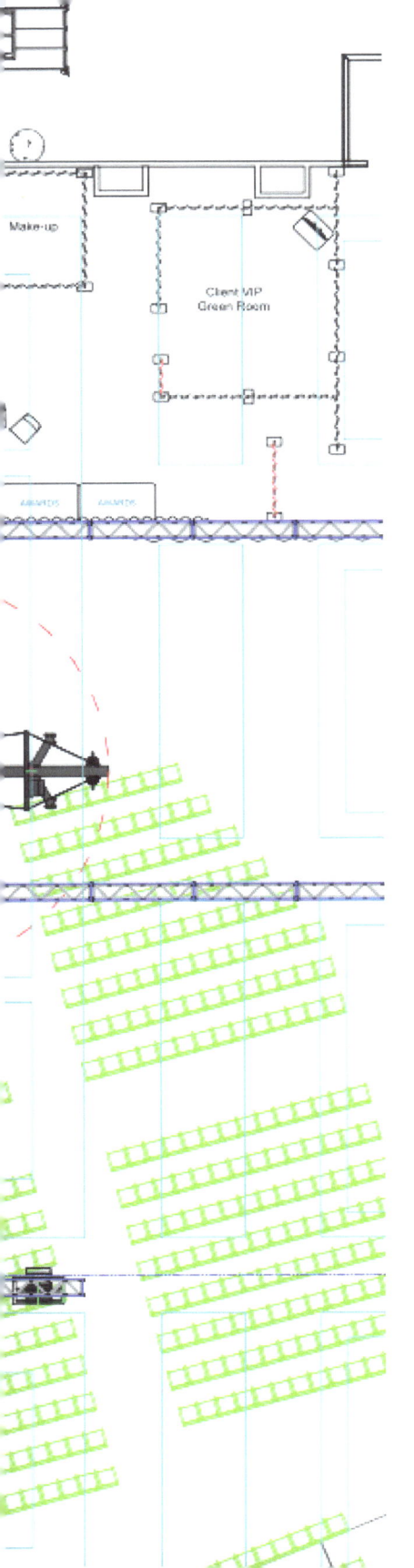

Table of Contents

1	What is a Video Projectionist?	20
2	Start with Ownership	29
3	Requesting Documents	39
4	Identify the Venue	48
5	Identify the Show and the Client	58
6	Know the Different Roles	65
7	Understanding Your Drawings	73
8	Understanding Your Equipment	84
9	Understanding the Production Schedule	91
10	Teaching Others to Project	98

1 | What is a Video Projectionist?

Imagine you are attending a conference, meeting, music festival, or any kind of large event. There's an audience, screen, and presenter wanting to convey a message. From a PowerPoint or Keynote presentation, to architectural or medical designs; from charts and graphs, to an artist performing on a stage, the content that is magnified and displayed on the screen is often produced by a video projector. Behind that projector is one person managing a team. That person ensures the image shows up clearly aligned, color balanced, and properly converged across multiple images. **That person is the Video Projectionist.**

By walking into your local electronics store, one would think the job was as simple as connecting a computer to a projector and voilà. You're done. However, it's so much more. A Video Projectionist has to consider projector placement (and rigging if necessary), signal flow, power consumption, cable paths, light output, screen surface type, aesthetics, and a myriad of other factors to ensure the projected image meets the client's approval. Another component that is also important to the Video Projectionist is the video source. Whether it is a computer, media server, high resolution switcher, or Blu-ray/DVD player, this piece of equipment can determine the complexity of the setup.

For the purposes of this textbook, we will approach the role of the Video Projectionist from the perspective of executing a large-scale corporate event or convention. We will explore Video Village and understand how it plays a part in our connectivity. We will also discuss how various crew members play their part in facilitating a successful load-in. The Technical Director, being one of them, translates the client's vision into a drawing, or layout, that shows the room in a grid-like display with the equipment, staging, rigging, and chairs for attendees. The Video Projectionist works hand-in-hand with the Technical Director (TD) to figure out the cable path, weight distribution, line of sight, throw distance, etc. to ensure that the image on the screen looks beautiful from every angle.

However, before we dive deeply into the strategic plan of the execution, it is important to:

Know the History

If you understand the roots of video projection, you not only understand how it works, but also how early inventors improvised and why they created their inventions. Telling stories and communicating through imagery dates back to prehistoric times with the Homo habilis, however, image projection truly evolved in the 17th Century.

The Magic Lantern

In 1659, the Dutch scientist Christiaan Huygens invented a slide-based image projector that he dubbed the *Magic Lantern*. It consisted of ten small sketches of a skeleton taking off its skull. These images were illuminated from behind and projected onto a wall. The light source came from lanterns and oil lamps and passed through the slide by a

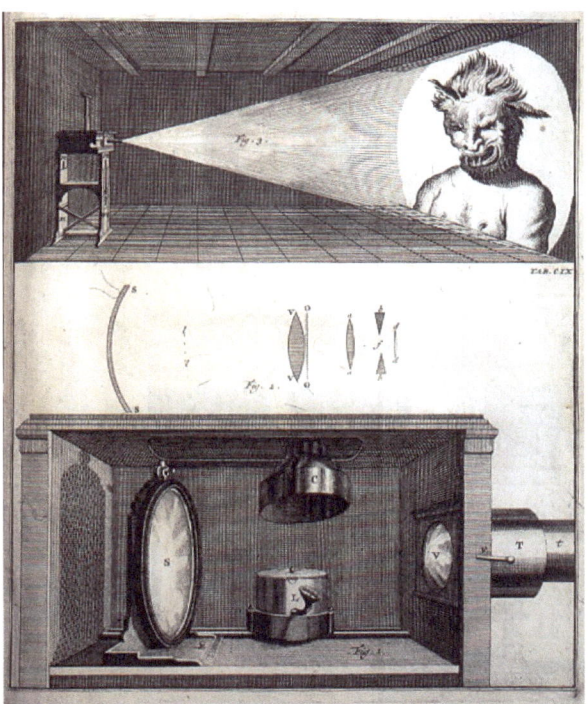

A page of Willem 's Gravesande's 1720 book *Physices Elementa Mathematica* with Jan van Musschenbroek's magic lantern projecting a monster.

concave mirror situated behind the light. Huygens's lantern used three lenses to focus the plane of the slide on the wall. Black paint was used on the slides to block out excess light, and to reduce distortion.

The Episcope

In 1756, Leonhard Euler, a Swiss mathematician, astronomer, and engineer, invented an opaque projector called the Episcope. This projection system used opaque materials to create the projected image by shining a bright lamp onto the object from above. Mirrors, prisms, and imaging lenses were then used to focus the image onto a screen. The concept was expanded upon by other inventors, including Henry Morton. Morton once used an oxyhydrogen lamp to project an image in the Philadelphia Opera House that was large enough to be seen by an audience of 3,500 people.

The Zoopraxiscope

The beginnings of movie projection originated with this invention by Eadweard Muybridge in 1879. The projector used 16" glass disks with painted silhouettes that were run in sequential

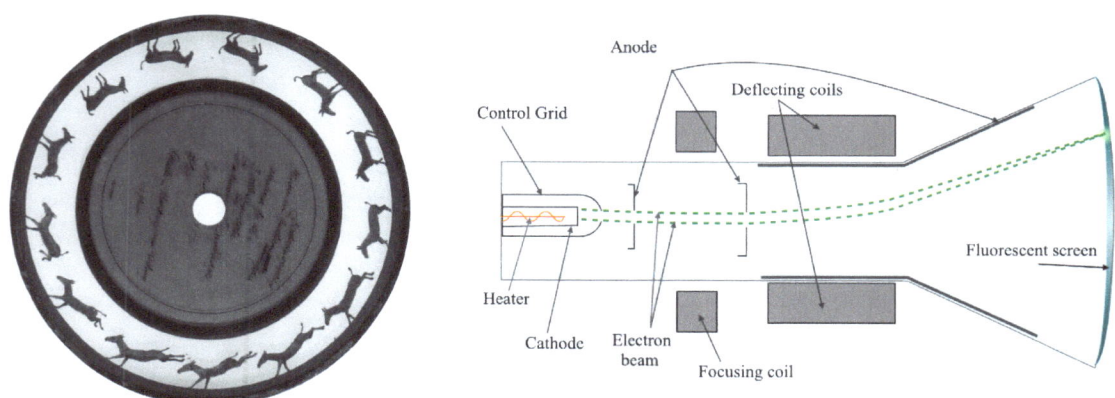

Left: Black-and-white picture of a coloured zoopraxiscope disc, circa 1893 by Eadweard Muybridge and Erwin F. Faber. **Right:** Cathode ray tube using electromagnetic focus and deflection

order. The images were elongated to compensate for the distortion of the projection. With a hand crank, Muybridge could roll the film and create a moving image. A few years later, Leon Bouly took that technology and evolved it into the cinematograph: a camera, projector, and printer in one. At the Paris Exposition in 1895, the Lumiere brothers projected a short film on a screen that was 16 x 21 meters, or approximately 52.5 x 69 feet.

Overhead Projectors

In the 1950s, overhead projectors capitalized on the technology of the 35-mm slide projector (which evolved from the Magic Lantern) but were able to use much larger transparencies than the previous slides. An overhead projector used a focusing lens to project light from an illuminated slide onto a screen or wall. Because the transparencies were much bigger and often used for lectures, they had to be placed face-up to be readable by the presenter. A mirror was placed inside the projector just before or after the focusing lens to reverse the image and allow it to project accurately on the surface. 35-mm slide projectors and film projectors do not use mirrors because the slide's image is non-reversed on the side opposite the focusing lens.

Cathode Ray Tubes

During the same time period, Cathode Ray Tubes were invented. These are vacuum tubes that contain one or more electron guns. The electron beams are displayed on a phosphorescent screen. Color images required three different electron beams in the colors of red, green, and blue. These systems were called three-gun projectors and were complicated to converge, because all three CRTs had to be precisely focused to prevent blurring.

My mentors used CRT technology, and when I first started in the industry, I learned the basics of the three-gun alignment method. Even though we weren't using CRTs then, we were doing similar processes by aligning primary and backup projectors to increase brightness and minimize the risk of a non-image portion of a presentation. That understanding of three-gun projecting expanded into my process of aligning multiple projectors and converging widescreen blended images.

Digital Light Processing

Developed in 1987 by Larry Hornbeck of Texas Instruments, this optical micro-electro-technology uses a digital micromirror device to project its image. With access to better technology, DLP displays had better light usage than previous projectors. This resulted in clearer, brighter, and sharper images. Digital technology also meant multiple color processing, less jittery images, easily replaceable light sources, and much lighter display units. DLP projectors were more affordable and had much higher resolution than any other display technology on the market at the time. In fact, *Star Wars Episode 1—The Phantom Menace* was the first movie distributed with this form of processing in 1999.

Before the advent of digital technology, Video Projectionists had to be inventive to increase the image's size, clarity, and color quality. In many ballrooms, the projectionist would stack multiple slide carousels on top of one another, with the same slide show in each, and rotate all of them at the same time. It was a complicated system, with a lot of room for error.

Today's Video Projection

Then computers and digital imagery came along. Although they have revolutionized the industry, they have not taken the place of a good Video Projectionist. Computers are there to make your job easier, but they aren't meant for you to take shortcuts and compromise the quality of your product. As a Video Projectionist, it's essential that you understand how to manipulate and align your projector so that you can maintain the integrity of the pixels and light when the image is magnified over the throw distance.

I was fortunate to learn from some of those first-generation, three-gun Video Projectionists, and I saw the value in what they taught me, what they saw, and what they pioneered. Their inventive spirit is my guiding light when I am on a job and something breaks or doesn't work properly, and I need to improvise. Video Projection is an art and a craft. You have to respect its history in order to be successful in the present day and future.

The job of a Video Projectionist can, at times, be very complex. At any given event, I can have anywhere from two to twenty projectors. That means I must white balance every image and vertically and horizontally align their grids so that all images are consistent no matter where you are sitting in the room. This is paramount when working high-end corporate events. The branded colors of a company's logo must be spot on, and the presentation must reflect the client's vision.

Knowing, understanding, and having the ability to translate the client's imagination into a tangible reality is a critical part of this job. Compare it to the way an interior decorator works. A client comes to the decorator wanting a cozy space or a modern feel, and the decorator carefully chooses everything from the wall color to the paperweight on the desk to match that vision. They hire people to install the molding, paint the walls, and construct the bookcases as part of the execution. That's what we're doing as Video Projectionists. We are helping to paint the wall, and we are using our skills to help tell a story.

My first responsibility for every event is to understand that story. In order to deliver a quality product, I must first see my client's dream. They have a vision of their conference, concert, or event, and have hired me to execute it in a technical way. I have had the opportunity to work with Fortune 100 companies because of the love, care, and pride I put into my job. I do my best to represent their brand no matter the period of my contract. Whether it's for a day or a week, I put my all into every gig, and that's why I get paid what I do.

Responsibilities and Job Duties

A Video Projectionist must understand some key responsibilities prior to arriving on-site:

1 | ASK QUESTIONS

One of the things I covered in Act One of my book *Career Projection 101* was the process of understanding and executing the client's vision. That person has an idea in their head, and it's up to you to politely and patiently extract it. Then, you have to take that information and understand what they are saying. Sometimes, you have to take notes and keep asking questions until you have a clear-cut understanding of the job expectations, the venue, the equipment and crew needed, and the time allotted to complete the task.

2 | MANAGE THE CREW

At any large event, you have a crew of people working with and under you. They're going to be looking to you for leadership and answers. You must have those answers, or at least the wherewithal to get them.

3 | LEAD AND DELEGATE

You can't be everywhere at once, which is why hiring a competent crew is important. You have to trust who you are working with and empower them by delegating responsibilities. Provide them with the right amount of information to do their job, then let them do it.

4 | UNDERSTAND THE EQUIPMENT

What equipment are you being provided, and what equipment will you need to bring in, to execute what the client expects and wants? If the rental company or venue is providing a lens, it's your responsibility to know if the image will travel the throw distance and reach the screen. It's also your responsibility to know if you have enough cable to go from where the projectors are located to the signal source. These are just a couple of examples, but it's important to know what you need and what you have.

5 | BE A TEAM PLAYER

I have to work with Rigging, Audio, Lighting, Scenic, and Stagehands on any given project. Sometimes I have a Projection Assist working with me, and sometimes I don't. Either way, I have to be a team player with all departments on a production. This also includes the Graphics, Creative, and Stage Management teams. We are all working together to bring the client's vision to life.

6 | BE A PROBLEM SOLVER

I often reflect on a job I did where the projector had a fixed lens. I struggled to make the image fill the screen, because I didn't know how to properly use the equipment. The fix was simple, and all I had to do was ask for help. Lesson learned. Don't be afraid to look stupid in the moment in order to look professional in the long run. It's okay to say, "I don't know, but I'll find out."

7 | MONDAY MORNING QUARTERBACK

Your role as a Video Projectionist is collaborative. If something goes wrong or doesn't come off as well as you would have liked, you have to tap into your resources, trust other people's knowledge, and be willing to try and fail forward so you can succeed in the long run. At the end of every job, analyze what went well and what didn't, so that you can improve on those areas and continue to succeed.

In today's high definition world, there is even more pressure to create "perfect" images. To me, the job is like a Rubik's Cube. I can look at a grid, see the different corners and points, and know exactly what I need to do to manipulate the image and make the viewing experience pleasant for the audience. This has developed into a life skill I call Projection101.

The industry is growing and changing every day, and we have to grow with it. Projection mapping is becoming more and more prevalent, and we need to understand light geometrically to adjust our image to fit the unique shapes onto which we are projecting. Yes, there are computer programs to do this, but a good Video Projectionist knows how light works, and knows how to use the tools at hand to make an image look its absolute best.

What is a Video Projectionist?

For Video Projectionists who truly understand the manipulation of light, there's an ever-growing market of opportunity available as this form of art, publicity, and branding becomes a part of every industry. The opportunities are there. You must be ready to grab them.

> **Good geometry prevents a multitude of sins. The critical part is getting your image square. Get your geometry right, then you won't need to warp your image unnecessarily. Resist the urge to use the warp as a crutch. It should be used last and sparingly.**
>
> **At the same time, don't let perfect get in the way of good. You have to know when you've done a good job. If your client is happy, don't waste everyone's time trying to make your image 'perfect'. I will never be.**
>
> Danny Harris | 40 Year Video Projectionist | Dallas, TX

2 | Start with Ownership

When I was in the fifth grade, I joined my elementary school band. I was a member of the brass section, and I played the trumpet. Everyone in the band, from percussion to wind instruments, worked in unison to produce a certain sound. We each had our part and our section, and we came together to ensure the audience's experience was the best it could be.

That's how it works in the production industry as well. **The baseline of everything we do as Video Projectionist starts with accepting our part**, and understanding how our role plays into the event at large. Knowing the equipment is one component, but knowing how and why your role is important is another.

The Role of Clarification

The main role of a Video Projectionist is to clarify and magnify the message. The client has a story they want to tell, whether it's for an annual meeting or a concert, and your job is to take what they are trying to say, and deliver it in the best and clearest way possible. Being part of the Video Projection team means you are part of the message, part of the display, and part of the event. When the audience looks at that screen, grid, grayscale, color balance, alignment, etc., it should all be a representation of the love, care, and detail you put into your craft. In essence, a reflection of you and your pride in your work.

As a team, we all have to work together (similar to how I did in band) with the client and the presenter to deliver a message that is clear, concise, and uninterrupted. If my projectors are out of convergence, or the colors are off, or the screens are flickering because of a bad cable, then my section is a distraction and is taking away from the message and the experience.

To make sure that doesn't happen, my first step is to ask questions and go through my checklist prior to arriving on-site. I want to understand different elements of the production, even ones that aren't in my department. The more information I have, the better I can do my job, and the better I can serve my client. It's not just about asking equipment questions, either. I need to know the crew, the production schedule, and the layout of the room. All of this information helps to create a safe and positive work environment. Through my years of experience, I realize questions give you clarity, and clarity ensures we all do our best.

Take Ownership from Day One

I remember working an event in Chicago and overhearing a Breakout Technician tell the client, "That's not my job…you need to call someone else to take care of that". Hearing their tone and the way they addressed the client, I thought to myself, *They don't get it. They don't realize this is a team effort, and we are all in this together*. That Breakout Tech only saw their duties on their To Do list. They didn't take the initiative to handle the situation themself or think about serving the client to the best of their ability.

Taking ownership means having pride in everything you do and doing all you can to help create a memorable experience. Whether you are a Stagehand or a Projection Assist, taking ownership includes, but is not limited to:

- Running cables and making sure they are neatly labeled
- Having the tools you need to do the job, but if you don't have them, buying them prior

Start with Ownership

to arriving on-site (Stagehand - https://www.clemcoav.com/stagehand101/ and Workbox - https://clemcoav.com/workbox101/)

- Learning as much as you can about other departments so you can communicate across disciplines, and so you can understand how everything comes together
- Doing a post-event analysis and asking yourself, "What lessons did I learn working on this job? How can I improve my skills and performance? What will I need on the next gig that I didn't have on this one?"

I'm always trying to think forward. "What can I do to make my job better? How can I provide a better experience for the client, the end-client, and the other crew members? How can I make today's load-in or load-out better than the previous? Lastly, how can I be a better team player?" Answering those questions, and thinking ahead to improving the next project, is a part of my role. Moreover, answering those questions, and taking the necessary actions to improve myself, is a part of **owning** my role.

Be Ready to Learn

When you first start accepting gigs as a freelancer, you will quickly realize that you aren't always given the resources you need to be successful at your job. Whether it's tools or information, you are often lacking something. In the beginning of your career, it will be information. However, you won't realize it because you won't yet have the experience to do so. Now the question is, "How do you know what you don't know?"

The first step is to humble yourself. Many technicians, including myself, start off with an extreme confidence in a particular area. What

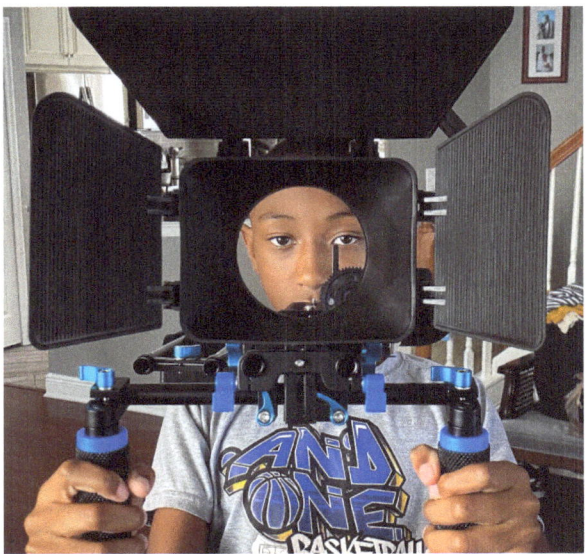

Looking through the lens, and now knowing what I know, I realize we all aren't given, or can afford, the tools to succeed. However, if you make do with what you have, and believe in yourself, you can make the impossible possible. See it... Believe it... Achieve it! #Projection101

can be perceived as arrogance, can get in the way of establishing relationships and learning from others around you. Pay attention, ask questions, and be eager to learn from those who have come before you. I quickly realized that I needed to look at what I was doing as an apprenticeship journey. I wanted to genuinely build relationships with people who had gone through the fires and trials, so that I could become better through their experience. When I was thrown into jobs on my own and was in over my head, I tapped into the relationships I had built. From Projectionists, Riggers, and Engineers, they were willing to share their advice and support because I was willing to learn.

When I first started in this industry, I struggled finding the people and relationships I could lean on. I was out in the field doing my own thing and often felt like an imposter. I didn't know everything I was expected to know, and I didn't have someone I could be vulnerable with and trust to walk me through the steps towards my success. I didn't have a mentor, and I was in desperate need of a teacher. I needed someone who could watch and coach me along the way. Similar to a person in pilot school, I needed an instructor to teach me prior to taking my first solo flight. Someone to fly alongside me and make sure I didn't crash and burn.

As I grew in my career and built those relationships, I realized my mentors became more successful as I became more successful. They grew in the industry, whether in the field or in the office, and, to this day, revel in my accomplishments. They know that just as they mentored me, I will do the same and become a mentor to others. It's like a second generation/second layer in ownership. For me, knowing people I helped are becoming successful and doing well gives me an enormous sense of pride. They are an extension of me, just as I am an extension of others, and I am thrilled to see them do their job confidently.

The people who helped aligned my path to become the man and Video Projectionist I am today will never know how grateful I am for their guidance. All I can do is continue to pay it forward and backward, helping to provide jobs and opportunities, and then shining light on the brilliant people I have worked with along the way. That's what I do with the Production Channel podcast. I interview some of the best and brightest in this industry, allow them to share their journey,

Start with Ownership

and talk about the people who mentored them along the way. Similar to Kanye West's "Roses", I believe in giving credit and thanks in the moment and not waiting until a particular date or occasion.

When Things Go Wrong

Mistakes happen. They happen to all of us at one time or another. So when they do, own your errors and faults. The people who survive in this industry, the people who make it and are successful, are the ones who can admit their mistakes and do everything in their power to make sure that mistake doesn't happen again.

I learned that lesson early on. On one of my first gigs as the Lead Video Projectionist, I had to do an overnight, half day load-in at the Marriott World Center in Orlando, Florida. The setup was a ten-foot scaff tower, a single projector on top, and a 16 x 9-foot rear projection screen. As I mentioned in the previous chapter, I had trouble getting the image to fill the screen. I did everything—went through the menu settings, checked the lens for any noticeable issues, and looked to see if there were any internal wire or connection issues in the projector. Nothing! I couldn't get it fixed. I didn't think to search the internet (and back then, the internet wasn't as full of information as it is now), nor had I established relationships with other Video Projectionists that I could just call and ask for help.

The art of taking a basketball arena and creating an all-encompassing banquet hall starts with finding your perfect convergence. Use patience, peace, and understanding as a guiding light to overcome any obstacle. #Projection101

After the scheduled 5-hour shift ended, and the rest of the crew left, I told the Technical Director and Video Department Lead I was going to work off the clock and stay there until I figured it out. I worked all night, and into the next morning. At 7:00am, I saw the Video Projectionist, who was managing the general session next door, enter his ballroom and turn on his projectors to begin warming them up. I knew he was working for another company, but I had a ticking deadline and a problem I couldn't solve. I asked for his help.

Within a few minutes, he figured out that my lens was a .8, or fixed, lens. That meant it didn't zoom. I had no clue that the numbers on the side of the lens referred to the optical zoom ratio and how tight the image could push in or how far it could pull out. Something so simple, yet something I was never taught or knew to look up. As this seasoned Projectionist explained this information to me, it began to make sense. That would explain why every time I tried to zoom the image out to increase its size, it wouldn't work. The fix was to move the projector further away from the screen. By pushing the scaff tower back a couple of feet, I was able to "zoom" out the image, and have it done in time for the event.

By humbling myself and being vulnerable enough to ask for help, I was successful in the long run. Yes, that momentary embarrassment was gut-wrenching and humiliating, but by accepting my faults, the presenter was able to display their presentation the way it was intended. The client was happy, and I learned the value of building relationships with other people in this industry.

Not everyone will be eager to help you. Unfortunately, some people are threatened by new faces and are hesitant to share their knowledge. This is often because they don't want to lose their upper hand or help others climb the ladder to success. Don't let the challenge of finding the right people who are willing to share and teach deter you—keep looking. They are a resource that will help you navigate your path, and it is worth the effort to build and maintain those bonds. I know for a fact that I wouldn't be where I am today if someone else didn't take ownership of my success and wasn't willing to coach me along the way.

Always Have Respect

When you're new to your position or new to a team, it's easy to feel intimidated by the people above you. Trust me, I'm saying this from first-hand experience. I've been on jobs where I noticed things that seemed to be wrong, like a cable run backwards or an error in a math calculation, and I was hesitant to speak up. However, I knew if I didn't, it could impact the entire show.

My strategy to address the concern was to ask for clarification in an inquisitive, non-confrontational way. As a newcomer, you don't want to come off as arrogant and say, "You did this wrong!" That may not be received well by the veterans on the team. Instead, approach it from a place of humility. "I'm curious why you did it this way…"

The people involved may have years of experience on you, and could have intentionally changed the way things are done because they know it will suit that particular situation. In this industry, we're always adapting and changing to fit different clients, venues, technologies, and scenarios to yield favorable results. If you go into the situation looking to gain understanding, it shows you are trying to be part of the solution. When you do catch an error, hopefully the other crew member acknowledges your catch. In doing so, it will help you build your confidence in that particular area.

FCT

Whatever you are doing, whether it is executing a job, asking for help, or building your network, it all boils down to FCT:

- Familiarity
- Comfort
- Trust

This is something Reggie Butler taught me years ago. If you have those three things with someone, then questions and conversations aren't viewed as confrontational. They are viewed as a mutual respect of being in this together to exercise the clients' vision. Now that I have decades of experience in the industry, and understand this from a management perspective, I try to help others see the importance of building relationships and rapport. Once they have been established, that bond can help to eliminate doubts or uncertainties and create a harmonious environment of grace. We all make mistakes from time to time, but when FCT is present there's empathy and room for people to take ownership of their actions.

What do you do when there's no one there to help you during an overnight load-in, or an early morning event? What do you do when there's no one there you know or trust to help with your video projection issues? You turn to the Video Projectionist Support Group.

VPSG is a private Facebook group I created to provide a twenty-four hours a day, seven days a week resource for Video Projectionists around the world. The technology is ever changing, and we increasingly need to rely on other experts for help. I curated this group, and interviewed every applicant, to ensure they were willing to be of assistance to others. I wanted a community of people willing to help, and that's what VPSG has become over the years.

Know and Live Ubuntu

There is a Zulu word that Nelson Mandela often used to inspire the people of South Africa: *Ubuntu*. The most common translation for it is *I am because we are*. It is a philosophy for teamwork at a different level, one where everyone is being the best they can, for the greater good. If you happen to catch the interview with Coach Doc Rivers on the Netflix show, *Playbook*, you'll hear him talk about how he and the Boston Celtics lived by the Ubuntu philosophy, in good times and bad.

Ubuntu recognizes and celebrates each person's uniqueness, their skills, their talents, and sees them as part of a universal effort. It's about belonging to every person around you, and them belonging to you, so that you are collectively working to make the world a better place. Then by extension, we are all giving others grace in their humanity.

Ironically, Doc Rivers was the head coach of the Orlando Magic when I first started working there. This philosophy took root in his coaching a few years later, but I have believed in Ubuntu from my first day in production. You have to. Many of us travel to work events, and on those trips, we inevitably build relationships and friendships with one another. Sometimes our home lives suffer, and sometimes we take our own frustrations out on people at work.

When you come from a place of Ubuntu, you look at each other as not just a fellow technician or team member, but more so a brother or sister. With Ubuntu, your compassion for one another is immeasurable. "Where you may have fallen short, I will take on your load, and where I am falling short, I pray you will be my support."

When I was in college and pledged my fraternity, we had a ritual where we carried our line brothers on our backs and ran across campus chanting, "He's my brother; he's not heavy!" This taught us to support one another, and that the load we carry is not our own. It is bearable because my brother is there with me.

When we work an event, we all have a common goal to accomplish. We are going to do it together, and we should all look to do it with the best possible attitude. None of us can take this journey alone, and the more we support each other, the more we are supported in turn.

No one in the audience knows the amount of work that went into creating the masterpiece they are experiencing. They can't comprehend the hours of prep and labor that was put into that one event for that one client, and they don't realize it was a massive team effort. There's no way I could walk into a ballroom and build a complete show from audio, video, staging, rigging, and lighting by myself. Nor would I want to. We are all dependent upon one another for our individual success.

Glenn "Doc" Rivers is a former National Basketball Association (NBA) player and current head basketball coach. Doc is a 3x All-Star head coach, was awarded Coach of the year in 2000, and won a Championship title with the Boston Celtics in 2008.

Photo by Jason Miller, Getty Images; https://www.theringer.com/2020/6/16/21292988/doc-rivers-on-player-empowerment-growing-up-son-of-chicago-cop-where-country-is-going-and-nba-return

Start with Ownership

When you understand that, you will approach and treat your job differently. People will then approach and treat you differently because they know you are going to do what it takes to help others succeed. You will find more satisfaction and pride in your craft, and every day you work, you will feel blessed.

When you are in an industry that you love, and you feel passionate about your work, you will naturally execute everything to the best of your ability. When I got into this field, I had no idea how much money I could make. I was here because I was excited to use the gifts and talents God gave me. Because I was focused only on doing my best, serving others, and taking ownership of what mattered, the money followed. With every single job, I put my heart and soul into the work because the image on the screen is just as much a reflection of my love for this industry as it is for my client.

Plain and simple, sometimes you just have to take the ownership because the indirect approach doesn't always work.

When you know what you know, and are confident in your ability, your race, gender, and nationality doesn't matter. It's about getting the job done. Who's the best person for the job? Make them your Lieutenant. Who can take orders but will also lead? Make them your Sergeant. Who will get down there and do the work at all costs? Those are your Privates.

We all play a role, and when your squad owns the mission, we get it done!

Melvin LeGrand | 30 Year Video Projectionist | Phoenix, AZ

3 | Requesting Documents

As we discussed in Chapter Two, being a successful Video Projectionist starts with taking ownership of the job from beginning to end. **That means doing your pre-production work prior to arriving on-site**, so that you're ready to build and execute your event. There's always a chance that something could go wrong or not go as planned. To reduce the number of potential risks, I follow a motto known as the 6Ps—Prior Planning Prevents Piss-Poor Performance. If you live by this adage, you will have a greater chance of success in this industry.

When a client calls and asks if I'm available to work a show, my mind is engaged and begins prioritizing everything I need from the moment I say yes. From documents to travel arrangements, every facet needs to be thought out and accounted for. In my experience, clients rarely provide drawings and equipment lists without being prompted; it's my responsibility to contact the right person and request the necessary materials.

I have a proven method that has allowed me, and so many others, to be successful as a Video Projectionist. Part of my process is analyzing and planning. I must be fully prepared the day I walk into the ballroom, and I want to ensure I have everything I need to do my job successfully. If I don't take my pre-production time, then I'm putting the show at risk. It's important for me to understand, and over-stand, every element of the content displaying process so I can execute my role effectively and efficiently.

The Drawings

The layout of the room, whether it's provided by the Technical Director in a Vectorworks, AutoCAD, or PDF file, will help you understand what you are building, and what your portion of the build will consist of. Think of it like a blueprint for a home. The General Contractor helps oversee the full design, but then calls in different trades/independent contractors to work on their parts, such as plumbing and electrical. The part we're obviously focused on is video projection, but there are several other trades in various departments as well.

On a drawing, you can see how many screens you're going to have, whether they are front or rear projection, how far your throw distance will be to the screens, and what cable paths you will need to take. For example, if the distance from Video Village to your furthest projector is 350 feet, you'll need to plan for that distance because we don't typically run power further than 250 feet,

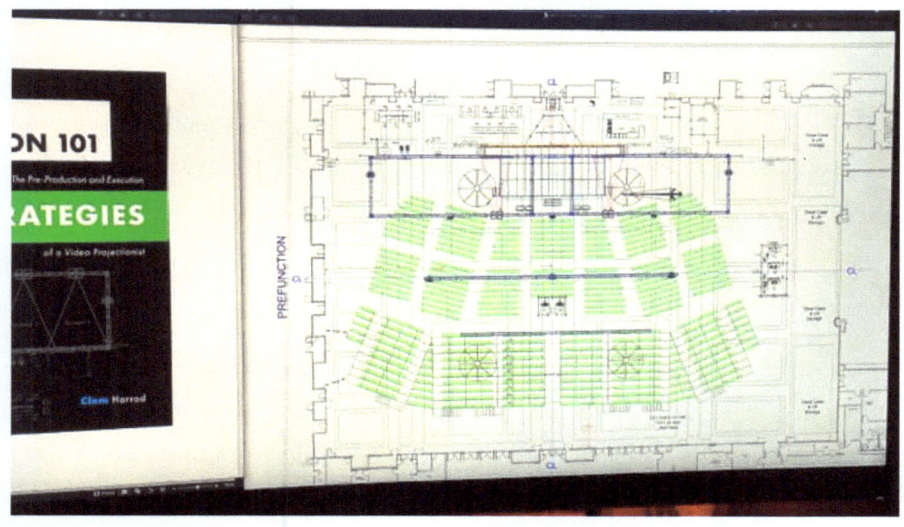

What you begin to understand from looking at a drawing is there is life happening and things moving within the 2-Dimensional space on the paper. In my mind, I can see the people working, I can feel the cases being pushed, I can hear the scissor lift beeping, and I can react to the Rigger yelling, "Truss coming in!"

due to voltage droppage. When the drawing files are provided in Vectorworks or AutoCAD, you can measure your runs in the program, plan accordingly, and take notes to recall when on-site.

It's a great benefit if you have previously worked in a particular venue. You'll know the best route to the loading dock, the orientation of the ballroom, the location of the rigging points, etc. But if you haven't, you will need to study and dissect the drawings, and more importantly, ask questions. You don't want to show up on the day of the event, trying to understand what you're building and whether you have all the equipment you need. That's a recipe for disaster. I know, because I did that early in my career, and I learned my lessons the hard way.

The drawing analysis is one of my favorite parts of the job. When I was a kid, I loved architecture. Whenever I moved into a new apartment or when my wife and I built our first home, I looked at and analyzed the drawings. I could look at them and see how the furniture should be oriented to maximize every square foot, and what our lives would be like in that space. I use those same visualization skills when I am looking at a drawing for an event.

When you're new to this field, however, the drawings can be overwhelming and intimidating. That's where that mentor connection comes in handy. If you have made some industry contacts, volunteered at a show, or taken an on-site practicum workshop, you've had a great opportunity to go over the drawings with someone who has experience. There are also some virtual reality simulators available that can convert those drawings into 3-D renderings. It's really cool to see people on the stage, attendees in the audience, and images on the screen. The easiest way to practice these visualization and planning skills is to compare a home's blueprint to a real estate 360-degree virtual tour. It will help you see more than the flat lines on the drawing. You will begin to see actual rooms and spaces.

Another opportunity to increase your spatial awareness and learn from other professionals is by contacting our Workshops (https://www.clemcoav.com/workshops/) to shadow one of our Video Projectionists on a load-in. You will get a copy of the drawings in addition to other resources, and you will have an opportunity to sit back and see everything come together in the space. Our Video Projection 101 workshops have been an invaluable part of many technicians' careers. Our classes allow students to see the process, ask questions, increase their knowledge base, and begin to understand how the Video Projectionist's mind works.

My process includes a list of questions (which I explain in this book) that help me confidently check all the boxes when I'm planning an event. Some of those questions include: the name of the show, client, venue, layout, number of screens/projectors, rear or front projection, throw distance, cable path, and power drop location. The answers help me do my absolute best, and I know they will help you do the same.

Keep in mind, however, you may not receive all the documents, or answers you seek, the first time you ask for them. Don't be discouraged,

Workshops are personalized experiences that are tailored toward your style of learning and are meant to provide you with a deeper sense of why.

and don't be afraid to follow up. Politely ask again, and let the client know you are only trying to prepare and make their show a success. I often say, "This is part of my process to provide you with the highest level of service you deserve." That statement really puts it into perspective. I'm only asking so I can do my best for you.

The Equipment List

After you look at the blueprint and see what you are building, you need to ensure you have all the materials required. Walking on a jobsite ill-prepared for the task at hand is a surefire way to be noticed for the wrong reasons. You have to accept that it is your responsibility to be positive you have everything you need for the job. If you were a carpenter and showed up at a construction site with your hammer, saw, and tape measure, but not all of the required lumber, you couldn't do your job. The same applies here. Reviewing the equipment list shows you as a team player and lets the client know your plan is to execute a flawless setup.

Starting off, you may not know who to ask for these documents. I'd suggest you follow a chain of command. Start with your initial contact, which is typically the scheduling department. If they don't have the information, then they can point you in the direction of the Project Manager, Technical Director, or Account Executive who does.

Once you get the Equipment List, you'll see that it's full of everything needed to build the show. From my experience, much of that list won't pertain to you. Therefore, I recommend highlighting the things that do, and make a separate list of the additional equipment you will need to get the job done. Some of the more successful companies provide technicians with a standardized form to request unlisted gear. It isn't a guarantee you will receive everything on it, but your client will hopefully understand why you have asked and do their best to make it happen.

It all goes together with the drawings. If the "CAD", often a universal term whether it's created with AutoCAD, Vectorworks or PDF, shows that the projectors are 150 feet away from Video Village, but the cables on the Equipment List only allot for 50 feet, then there's obviously a

Imagine being on a jobsite of this magnitude and not having enough cables to make your connections. It could be very stressful and overwhelming. That's why it's important to have a detailed Equipment List and review your documents.

discrepancy. I recall working an event for the 2012 Republican National Convention in Tampa, Florida. Located at Tropicana Field, the home of the Tampa Bay Rays baseball team, the cable runs were extremely long. We had one fiber optic cable that ran from the Press Box, which was on the second floor above home plate, all the way down the third baseline to center field. Reviewing the Equipment List and Drawing prior to the event was a pivotal part of executing that build. Rick Wegner knew the 300-foot fiber cable wouldn't be long enough, so he requested the 600-foot reel that was at the warehouse. Because of the route we had to take around the seats and down different levels, the fiber optic cable was run completely off the reel. That was the first time I had ever seen that done, and I remember it because Rick respooled that cable at the end of the show.

Taking ownership of my role, requesting documents, and designing a process to recreate my work are some of the habits that have led to my success. When a company knows they can trust you to execute a show, and do it without drastically affecting their budget, you will be presented with more opportunities. Think about it. If my client is based out of Orlando, Florida, then any out-of-town show requires equipment to be shipped from their warehouse. That is an added cost of business. If I don't take time to review the Equipment List and Drawings, then there could be an additional cost to ship gear that is needed but wasn't initially quoted. By me reviewing the list, I'm telling the client, "I've got your back. I want to get this right the first time." That type of effort shows you care and are trustworthy. That type of effort gets you rehired because you care about the bottom line.

When you look at the Equipment List included in this book, you can see that it details how many projectors there are, what models we're using, what lenses they quoted, etc. You'll also have a list of cables and rigging hardware used for mounting or stacking the equipment.

In this scenario, the vendor is responsible for the major equipment, and I'm responsible for my own ancillary tools. I learned early in my career, from Projectionists like Stuart Brown and Melvin LeGrand, that you are only as good as your tools and your understanding of them. Since then, I began investing in my Workbox (https://clemcoav.com/workbox101/) and purchasing equipment that would help me repeatedly execute a successful build. From a standard and laser tape measure, laser level, crescent wrench, and power meter to a tactical pouch, lid organizer,

One thing I've realized from years of traveling on the road, is that no two workboxes are the same. From exterior stickers to the way the case is interiorly organized, each technician approaches their workbox in a different way.

https://www.youtube.com/watch?v=h1Tm8skxv4w&t=9s

utility knife, and Sharpie, my goal is to be prepared for any and all installation issues that may arise. By assisting and observing, doing the job myself for so many years, and always completing a post-show analysis, I have refined my process and improved on every single show.

The Production Schedule

This timeline goes from start to finish, from load-in to load-out, and gives you a precise idea of how much time you have to execute your portion of the build. The schedule gives you an intended flow of the day: from department call times, deliverable deadlines, and meal breaks, to client rehearsals, actual shows, and out times. When I'm reviewing the Production Schedule, I like to look for KPIs—key performance indicators—to set my pace. These are benchmarks the Technical Director (TD) uses to gauge the timeliness and quality of the work being done. For example, "all gear unloaded by 10 a.m." or "all projectors in place by end of day."

As I use a highlighter to mark the KPIs and other pertinent information to my department, I also take notes and look at how my schedule is impacted by other departments.

Then, I begin reverse engineering my steps. If all the equipment needs to be unloaded by 10 a.m., am I allotted enough time to do so? Based on the Drawing and Equipment List, how many projectors do I have, and how long will it take for my team and I to set them up? I am constantly thinking forward and backward, calculating the task and the time to make sure I can get it all done. If anything appears to hinder those goals, I have a conversation with the TD and keep them informed of my status.

Seeing the Production Schedule beforehand also helps me plan for staffing. What is my client asking me to do, and do I have enough help to accomplish it within the timeline? How many stagehands do I have? Do I have a Projection Assist? Have I worked with this crew before, or am I starting from scratch? After so many years of doing this job, I know what it takes for me to be successful. I also know how to build in any necessary buffers, or to ask for additional labor if I need it.

When you look at the Production Schedule, you shouldn't expect to leave the jobsite at 8 p.m. just because the schedule says you will. Especially on load-in days! You have to account for the variables and know that the schedule is an ideal estimate of time. Focusing on lunch or your out time when you first arrive can leave a bad taste in your client's mouth. If your heart is in your work and you're taking ownership of your role, then you understand that means staying until the work is done and you have planned accordingly. The schedule is based on a ten-hour day, but you should go into it expecting to be there for twenty-four. Things go wrong, deliveries arrive late, departments run into problems—build in that time to do the job right.

As you build relationships in this industry, you will have connections with Account Executives, Technical Directors, Project Managers, and Equipment Logistics Coordinators. By fostering these bonds, you will become privy to useful information that can help your show. Knowing what equipment is available, what is within budget, or if there are any room/setup changes, are all details that can help make your show its absolute best.

Even if you are working other shows in the weeks or months leading up to an event, make the time to do your Pre-Production. The more you do on the front end, the more organized and efficient you will be. Following these strategies and this process saves the client money, and in the end, makes everyone happier with the job you have done. You reduce the stress of a load-in because the team knows you have covered your bases. By building the show in your head and on paper prior to arriving on-site, you have already seen it. Therefore, you believe you can, and you will, achieve it. #Projection101

> **I did all that work upfront because I was kind of lazy. All the mental work you do upfront saves you physical work on-site.**
>
> **I always realized that requesting and reviewing my documents put me first in line to know where the power was and where my projectors were going. This saved me from some hard work and sweat.**
>
> **At the end of the day, I wanted to make my life in the room so much easier. I could do the paperwork upfront, or I could turn it into physical work on-site. I chose to do it as mental work upfront.**
>
> Steve Campbell | 19 Year Video Projectionist | Orlando, FL

4 | Identify the Venue

After confirming the event dates, one of the most important elements of any show is the venue. *The venue frames the entire experience*, and becomes the foundation of every decision you make, from equipment to setup.

For the past three years, I've worked a convention at Amelie Arena in Tampa, FL. This four-day meeting/concert is one of the largest non-sporting events held in this venue, and I have an advantage as a technician in this space. Since 2004, I have worked Tampa Bay Lightning games in this arena, and I know it very well. I have been in that arena so many times that I know exactly where the loading dock is located, which service elevators to use, and which floor to choose. In addition, I know where the projectors will live, and which paths the cables will take to reach them. That knowledge and experience helps reduce my pre-production time, and also helps me head off any problems that could arise. But what if you don't know the venue or you've never worked there before? Here are some of the most important elements to consider:

City and State

Knowing where the venue is located is a critical component for planning. If it's local, you can tour the ballroom in person and get a preliminary lay of the land. If it's not local, you are usually walking in cold and trusting the Technical Director's drawings. When I am working at a new-to-me site, I try to compare the unknown venue to ones I have worked in the past with a similar layout. This helps me know what questions to ask and what tools/resources I'll need to execute a successful build.

Equipment

For out-of-town gigs, chances are that not only are you traveling there as the Video Projectionist, but so too is the gear. Very few venues have a sizable warehouse on site, and most production companies have the equipment brought in by a rental and staging vendor. Similar to being able to trust the people on your team, production companies and TDs build relationships with rental and staging vendors to minimize the risk of equipment failures. If you know the provider and the people who work for them, trust they will deliver a quality product, and do whatever it takes to get the job done, then you will continue to use their services despite the cost to get the equipment on-site.

That being said, it is important to pay close attention to and review the equipment list for an out-of-town event. If the gear has to travel by freight or plane, you don't want to be stuck at a show site without the necessary lenses, cables, rigging hardware, or tools. It's not like you can run down to a local store and find a new projector. Cross-renting from a local vendor may be an option, but that's an added cost and added setup time for something that could have been caught prior to arriving on site. Knowing what you have to work with well in advance is not only necessary, but it's the responsible thing to do.

Staff

Does the venue have an on-staff Audio Visual or Rigging Department? Is it a union facility? Are we in a rural area outside of a major city? These are just a few factors that play a part in how the work will be done once on-site and the time it will take to set everything up. In short, local labor is key!

The quality of the local crew available to help you load-in, run a show, and load-out will take you from a profitable, stress-free event to an over-budget nightmare that will have you wanting to leave the industry. To understand the disparities in the workmanship, you have to realize that not all cities have a large convention labor force they can draw from for events. For instance, whenever I've worked at the Omni Grove Park Inn in Asheville, NC, we've used a labor crew that drives in from Charlotte to support our event. With a population difference of nearly 780,000 people, Charlotte is clearly a much larger market than Ashville and provides more opportunities for technical growth and understanding. Therefore, with a more experienced crew coming in, we have a greater chance of getting things completed on time and within budget.

After years of revisiting this venue, chances are the Stagehands and Projection Assists are people I have previously worked with. That means they know my process. Whether from reading my textbook, using my workbook, having taken an On-site Practicum Workshop, or going through numerous setups with me before, my team will be able to anticipate what I need. They become a valuable resource for me in pulling off the best possible event. *I am because we are – Ubuntu.*

Working with great companies in beautiful locations has been the highlight of my career. It's experieces like these that have opened my eyes to life's possibilties.

Conditions & Environment

Knowing where you are going helps you prepare for your trip not only as a technician, but also as a human. This is very important because your client needs and wants you at the top of your game. If they are paying you to execute and deliver your best, you must know what gets you there. Do you know what gets you to your peak performance level? Do you know your natural biorhythmic state? For me, it starts with a travel day and a mental reset.

I prefer building a travel day into my schedule and arriving at the venue the day before load-in, because it allows me to acknowledge and focus my energy in preparation of what's to come. I understand that getting acclimated to a city/state/environment helps me perform at my highest level, and I want to be centered and ready to lead my team. Therefore, I typically choose early or mid-morning departure flights, just in case I run into travel delays, and I also choose afternoon or late-day returns, just in case the load-out takes longer than expected. The last thing I want to do is leave my crew behind and working just because I have a flight to catch. It's part of my brand and responsibility to stay until the job is one hundred percent done. This includes getting everything packed up, put away, and sent back to the equipment provider either in the same condition or better than I received it.

Scheduling sufficient travel time has a major impact on your energy level and performance. If you're traveling all day to get to a load-in that night, you could possibly be tired and miss important aspects of your setup. This isn't something to take lightly. Safety should be on the forefront of your mind, and clouded thoughts and missed details could put others at risk. Doing your job well, executing to the best of your ability, and managing potential risks could be as simple as arriving the day before your call and getting a good night's sleep.

Knowing what city and state you are traveling to also helps you pack accordingly for the conditions. If I'm working outdoors or loading in a show, I bring shorts and a Dri-Fit collared shirt or a well-kept T-shirt. If I'm in a ballroom, which is typically kept around 65°F, I wear pants, a button-down shirt, and sometimes a sweater. Either way, I want to be sure everything I pack, from load-out clothes to gloves and hats, are practical yet professional.

Budget

While it's convenient and enjoyable to stay on site, especially if it's an extravagant hotel with a beautiful view, sometimes it's not possible. Larger conventions can take over an entire hotel and sometimes the traveling crew's lodging isn't within budget. Either way, this means you'll end up staying at an offsite location. When reviewing the reservation or checking into your accommodations, take into account any additional travel time needed to take your toolbox, gear, computers, etc., over to the venue. Budgeting your time and energy will help you prepare for the day's workload.

If you are given a per diem, or a budget for meals, parking, and other small conveniences, be sure to research the venue's pricing. Sometimes, the meals cost more than the per-day dollar amount you are allotted. This is something I've experienced staying at luxurious properties around the country. In these cases, a room with a kitchenette can be a valuable, money-saving resource if you buy groceries and snacks. In addition, the abundance and availability of food delivery services make it so you are not dependent on the restaurants at the hotel. You can order acording to your budget and increase your take-home pay.

One of my favorite things about identifying the venue where I'll be working is knowing which culture I will have an opportunity to become a part of. Besides traveling to Barcelona, the Bahamas, Lake Tahoe, Laguna Beach, and Washington D.C., to name a few, one of the best experiences I had was working in San Antonio, Texas. I stayed at the Emily Morgan Hotel while working a conference at the Henry B. Gonzalez Convention Center. My room overlooked the Alamo, and on the overcast morning of March 6th, I was awakened by very loud cannon fire. I was completely unaware, but pleasantly surprised, that I was in San Antonio at the same time as the Dawn at the Alamo anniversary reenactment event. The chances of that happening again were slim to none, so I seized the moment and stood outside to watch the ceremony. I love those opportunities to become one with a city and learn more about its heritage. I am able to expand my horizons and just appreciate life. When you travel, consider doing the same.

The Ballroom

Once you know where you are going, start looking at what you will be dealing with when you arrive.

Loading Dock

Where is it located? How far is it from the ballroom? Do you need to take a service elevator or navigate narrow hallways to get there? Is it a straight push or do obstacles abound? You don't have to wait until you arrive on-site to have these questions answered. Some venues have maps and other information available online. Do a quick Google search and become familiar with the property, its entry/exit points, its loading docks, and the other details that relate to your event.

Rigging Points

Every ballroom is structurally engineered differently from another. This includes rigging points. Some are located in the center of the room, and some are located throughout. Some venues use the high steel, and some use the airwall tracks. Some ballrooms, like the one in the Rio Las Vegas Convention Center, have a floating grid system that can be lowered so fixtures can be attached to it. Nonetheless, it is important to know what you are walking into before you arrive. Some venues' ceilings are too low to use hanging projectors because they will block the line of sight for the attendees. If you are aware of any obstacles a venue may have, it's good to discuss them with the Technical Director before the load-in so there is enough time to make necessary accommodations. Every ballroom, concert hall, stadium, and venue is different, and will offer different challenges. An open line of communication will help the team overcome them.

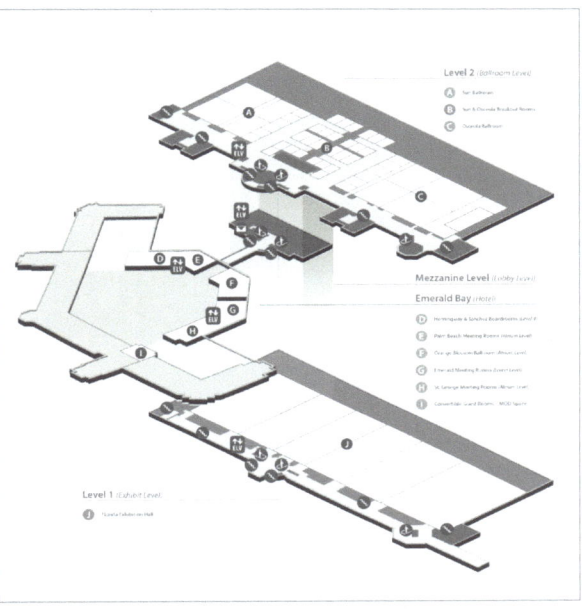

Using floorplans, like this one of the Gaylord Palms Resort and Convention Center in Orlando, Florida, can help plan the load-in/out of your event. Image by SIETAR-USA, https://www.pinterest.com/pin/403142604123177556/

A prime example is a show I worked at The Sanctuary at Kiawah Island in South Carolina. The ceilings were extremely low, and there were no rigging points to support our projectors. We considered building scaffolding towers for a rear, ground-supported setup; however, there wasn't enough backstage space. Therefore, we used truss towers, installed them vertically, and mounted the projectors on top. This way, there was adequate space for the attendees, the screens were large and visible, and the images were projected above the audience's heads.

The load-in and show at the Gaylord National Resort & Convention Center in Oxen Hill, Maryland went smoothly because of pre-production and planning. Photos by Steve Uhlmer.

Mouse Holes

There are times when all of your equipment isn't able to live, or fit, inside of the ballroom/venue. Mouse holes then are used to run cables between the walls and out to the equipment. Knowing where these are located makes a huge difference in how you will lay out your cables, and how long those cables will need to be.

I've typically used mouse holes when passing feeder cables from the back hallway into the ballroom. At the Mandalay Bay Convention Center in Las Vegas, Nevada, their power tie-ins are located in the service hallways. There, we have to use mouse holes to pass the feeder cable through because propping the doors open to the back hallway isn't preferred. Knowing and understanding these details about a venue can be useful when pre-planning for your event.

Power Drops

Power drops and power distros work together to energize the production. The power drop is the venue's dedicated power source used specifically for your event. This keeps your equipment on a separate circuit from the venue's other power sources. By isolating these circuits, your event can maintain its integrity and operate independently of other functions. For larger events, the venue provides multiple drops to further isolate the audio, video, and lighting department's individual power distros from one another.

The Basic Order of Power Flow

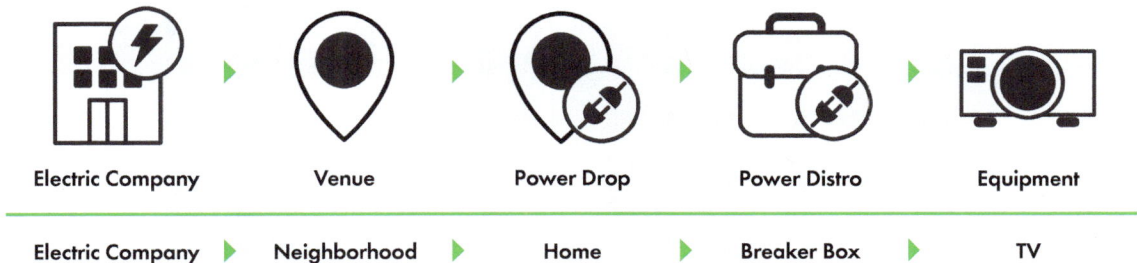

The power distro is essentially an enormous box-shaped power strip with multiple outlets that break into several circuits. Through this multi-breaker "power strip", each department is able to separate pieces of equipment to help balance the amperage draw per leg of power. This is why we, as Video Projectionists, keep our projectors on separate circuits from one another and from other equipment in the video department. It's to balance the loads, not overdraw one leg of power, and to not trip a breaker (more on this in Chapter 8).

A good way to visualize this is to think of a home in a neighborhood. The electric company generates the power that is available for multiple neighborhoods to draw on. Your neighborhood, being the venue, has a dedicated connection to the electric company. Within your neighborhood, there is a tie-in box that allows for your home, or event, to have a separate "power drop" from the other homes in the neighborhood. At your home, you have a breaker box, or "power distro", that isolates and allows a certain amount of distributed power to the outlets in the home. The outlets that are connected to the power distro are where the various pieces of equipment are plugged in to.

The living room of a home can be set up in a multitude of ways. The entertainment center could be on one wall with the seating on another, or it could be reversed based on your preference. It's really based on the experience you want to create, and how you want it to look and feel. When you have identified and know the venue, it is easier to explore and manage the possibilities.

That's where the Technical Director comes in. Based on their knowledge of the venue and conversation with its staff, the TD will note the power drops on the drawings, where the various departments will set up, and where the distros will live in those spaces. (Note: the distro placement can vary based on the location of the power drops and length of your available cables.) All venues differ. I've had some power drops on catwalks in the ceiling where the distro has either lived next to the drop or on the truss beside the projector. Some drops have been inside the ballroom and others have been in the hallway behind, with cables running through mouseholes. I've even worked in venues where there wasn't adequate power on site, and we brought in a gas-powered generator to run the entire event.

The key is to do as much homework as possible. As we talked about previously in this chapter, if you know what challenges lie ahead at a venue, you can sometimes see things the Technical Director or Project Manager have missed. That pre-production time allows you to support and help the entire event and not just you and your department. For example, if I'm looking at a drawing, I will know if the ceilings are low at the venue. I can then detect a line-of-sight issue between an audio speaker cluster and the location of my projectors. Doing the prework can save the team hours of on-site adjustments and strengthen already established relationships. It's about trusting and knowing we have each other's back.

There is limited time during a load-in to build trust from the ground up. This is ironic, because we have to construct our set this way, but not our relationships with people. You have to walk into the room with a certain amount of faith, establish an immediate bond, then execute to get the job done. When both you and the crew trust one another, you can work together on the mission, and believe in your ability to cut off any potential problems. Part of that relationship is on you, as a leader, to be prepared, know the answers before the questions are asked, and take detailed notes.

After the event is over, keep all of the documents/drawings, and store them in a file cabinet. Having mental and physical notes, as well as an organized process to recall the information, ensures success on future events.

Not all venues are created equal...

I quickly realized the loading dock and the egress to the ballroom can be the biggest obstacles. Long hallways, small elevators, tight turns, and food and beverage carts always seem to find a way to affect our push. There's no need to mention any specific places. We all know the ones...LOL

What you want is a standard amount of effort. You want to make sure the dock is clear, and that there's enough space to maneuver gear on and off the truck.

As we always say, 'Safety first!'

Steve Olson | 26 Year Video Projectionist | Phoenix, AZ

5 | Identify the Show and the Client

This might seem like an easy step: **What show are you going to be working, and what client will you be working for?** On the surface, it's cut and dried. However, there are many more factors that you should take into consideration during pre-production.

Do You Have History?

The first question you need to ask yourself is if you have a previous working relationship with the client or the show. Do you have history together?

For several years in a row, I traveled to Las Vegas, Nevada for a conference produced by the New York-based event marketing agency, Drury Design. When working events for this particular company, I know exactly what I'm getting into. It's going to be a massive, arena-style show with tens of thousands of attendees. It will involve several long days and nights of load-in, a lot of overtime, and an intricate show with phenomenal opening numbers. This means I'll need to be well-rested and ready to work when I arrive.

Over the years, my role with that show has evolved, and so have the challenges. As I worked my way up to Lead Projectionist, I paid attention to what the entire show encompassed. It was a constant learning experience, and as I took on more responsibilities, I had the knowledge I needed to ensure success. I knew what to ask for, what extra equipment I might need, and how best to prepare for the grueling hours.

Perhaps I'm working a small, high-end event for Orlando-based TEK Productions at a five-diamond resort near Lake Tahoe. Because I have worked that event before, I know the days are loose and fluid, I know the quality of image the client is expecting, and I know who I need to be to provide the level of service they require.

It can also work in the opposite direction. There are some shows I turn down because I know how disorganized the event will be. In a case like that, no matter how hard I work or how much I prepare, it won't be as successful as I want it to be. For me, it's better to turn down that opportunity than to deliver a subpar performance. When you are just starting out, you might not be able to afford to refuse work; as your career grows, you can be more selective.

Did They Bring in the Right Help?

With any event I have worked before, I run down a mental checklist of which technicians or production staff I have worked with. Some of these people I only get to see in person once a year, but I typically try to stay in contact with them via phone, email, or social media. By communicating with other technicians throughout the year, I am staying connected. Working together in person truly helps cement the relationship. Keeping relationships intact between shows is very important. Through this practice, I have built a reliable group to call upon when I need assistance, and they call upon me as well.

I LOVE my industry and enjoy every minute in the field with the team. We have the opportunity to build a piece of someone's imagination, and then allow others to come in and enjoy it. I believe it is something everyone should experience. (Left to right: Johnnie Dazulme, Kyle Prince, Omar Colom, Jonathan Draper, Clem Harrod, John Brewer)

When the production company knows the show and client, they start thinking about building their team (more on that and their different roles in the next chapter). If I haven't worked the event before, I research both the client and the show to get a feel for what the job will entail. A financial firm out of Canada and a tire manufacturer from Ohio will likely have very different events. One might be suit and tie in a ballroom, and one might be more casual in an exhibit hall. When a company considers the people they hire to work with them, they must also think about whether the crew is a good fit for that environment. One of the things we talked about in *Career Projection 101* was that not every person is meant to be with every client. Some people are great as roadies, others are happier in a more corporate environment.

The client and show will also dictate your expected appearance. I had a friend who worked the Clinton Global Initiative event in Harlem several years in a row. The first year, he had to buy a suit because the client expected all crew to be in a suit and tie. This isn't common in most markets, but you never know what a show may require. On the contrary, a music festival like Coachella will have a totally different vibe. Wherever you are going, it's important to match the image your client is trying to project. The more information you can gather ahead of time, the easier it is to match that image. The best option is to wear standard show black, so that you blend in with the crew and traditional backstage attire.

Production companies choose the people who fit the vision of the client, and whom they can depend upon. If this is your first job with this client, it's even more vital to do your research and know what they expect of you. For me, doing research consists of talking to other technicians, Googling the show and client, and searching social media for pictures from previous events. It's all about gathering data so I can provide that top level of service—from how I look to how I perform.

What is the Overall Expected Experience?

As an entry level Video Projectionist, you're typically not on the design end of the job. However, if you do your research on the client and the show, you will have some idea of what to expect for the feel and atmosphere of the event. If you haven't, you're going to walk in blind.

Identify the Show and the Client

This is why it is good to establish relationships with the Account Executive or Technical Director. You can ask them questions like: *What is the overall feel or mood for the show? What expectations do you have? What kind of experience do you want the attendees to have?* I know these are non-technical questions, and it seems like it's useless information, but every little bit can help you wrap your mind around the show.

Those conversations can also give you a preview of the work involved, so that you can preplan while you wait for the equipment list and the production schedule. In addition, asking questions like: *How many screens should I expect? Are we working with widescreens or traditional 16 x 9 screens? Is there any projection mapping involved?* can give you an idea of how to prepare. If you are booked far enough in advance, you can use your Projection 101 Workbook as a place to compile your notes, gather information, and become familiar with the event.

Through this process, it is important to respect other people's time just as you would want them to respect yours. While completing your workbook, you should pace yourself and gather the information in a way that you don't overwhelm the people you're working with. You don't want

I understand that executing a show with with quality-focused individuals like Keith Elliott and Mike Anderson of Watch Dog Technical Group will require numerous phone calls, emails, and pre- production time. Based on the scale of their events, I don't want to show up on-site without knowledge of what to expect.
Photo used with permission.

to bombard the TD, AE, LD, A1, or EIC with questions and insecurities. (We will discuss the various positions in the next chapter.) There's a fine line between asking questions to prepare for the job and asking questions and feeling confident you can handle the job, no matter what it is.

Overall, you should have a can-do spirit and mindset. The more you work in this field, the more confidence you develop in your abilities to problem solve, troubleshoot, tap into your network, and get a job done. I remember the first time I was hired as a Lead Projectionist. I wrote about that experience in *Career Projection 101*. Yes, I had an initial moment of panic—a moment of pause. But then I remembered that I had already been successful in so many other roles in my career. From Camera Utility to Score Bug Op to Stage Manager to Teleprompter Op to Chyron/Graphics Op to Audio Engineer; from Stagehand to Breakout Tech to Playback Op to Record Op to Camera Op, I had done a lot. I worked difficult shows, solved many technical issues, and had built a strong network to get it all done. I could rest assured that I had what it took to complete any task.

Problem solving is not just a great skill for working events, but for life in general. The more you apply the skills of analyzing data, understanding possible outcomes, measuring best results, and implementing the necessary course of action to all situations, the more successful you will become. Through this practice, you will gain the confidence to overcome any obstacles in your way. See it… Believe it… Achieve it! #Projection101

What If You're Completely New?

If you're just starting out and aren't finding the answers you need in your research? Now is the time to ask other technicians for advice. If you're just starting out and aren't finding the answers you need in your research, start asking other technicians for advice. I guarantee you will find the resources and support you need from fellow projectionists in the industry.

Similar to me, they keep show notes from previous events. In fact, I keep my drawings, equipment list, production schedules, and my setup details all on a template inside of my Projection 101 Workbook. In addition, I add a blank sheet of notebook paper for any memories and ideas that randomly pop into my head. This way, when I do my next event, or the next event with that same client, I have those notes as a reference. It's all about being prepared. The research you do ahead of time will help you on the pre-production calls and in the production meetings. Through taking this extra step, you will not only understand what others are talking about, but you will also be able to provide opinions and options.

After an event has concluded and the truck is all packed up, you shouldn't expect the client to call you and give you a rundown on your performance. It simply doesn't happen. Sometimes the only way to know if you did a good job is if they book you again. Even then, there's no guarantee. Your screens could have been perfectly aligned, the colors perfectly matched, the job executed without a flaw, but there could be something about you—your personality, wardrobe, or just the way you climbed the scaffold tower. Either way, something just didn't match the client's expectations. It's okay. Don't take it personally. Not every relationship is meant to be. If that happens, make a note for the next job. This way you learn from every experience.

I always thank my clients after the show and let them know I appreciate their business. I also ask if there was anything I could have done better, because I want to improve every single time. This may be via a phone call, email, or even over lunch. In the end, my goal is to make them happy and to continue to have that client as a source of income.

Sample Follow-Up Email with Client/TD

Dear [Client Name],

Thank you for allowing me to be a part of your event. It was a great experience, and I hope I provided a service that either met or exceeded your expectations. If I didn't, please feel free to let me know any areas where I can improve. I am always open to feedback.

I look forward to working with you again.

Sincerely,

[Your Name]

Identify the Show and the Client

When your client offers you constructive feedback, see it as what it is—a form of love. They aren't criticizing you because they don't care; they are giving you feedback because they want to see you succeed.

This is what I mean when I say Projection 101. Your present-time adjustments to the projector or the screen are all part of improving the future image. As a Video Projectionist, you get that. This mindset of improving your future image can also be used in your life. Over time, listening to feedback, learning from your mistakes, and learning from the mistakes of others are all ways to help you improve on your future work and your image.

Constructive criticism should be welcomed. It's not personal. It's professional.

> **History and an open dialogue are #1 to your success with a client. It also doesn't hurt when you have a stellar reputation that precedes you.**
>
> **I make sure every client gets the same show. It doesn't matter if you are a startup production company or the CEO of 'The Fruit Company'. I am consistent with every Jobs, and that ensures the clients always get the best that I've got.**
>
> **When you get ahead of every issue and excel at your role, you could be remembered as 'the best Projectionist in the country'. Humbly accept the title.**
>
> Stuart Brown | 40 Year Video Projectionist | Baltimore, MD

6 | Know the Different Roles

As previously mentioned, putting on a show requires a team of professionals operating as one. The strength of that team will not only determine the success of your event internally, but it will also determine how the attendees will engage with the message. **The key is understanding that no one person can do everything.** You must trust and empower your crew, then operate as a united front.

Knowing your part as one cog in the machine, as well as the role others play in the production, will allow things to run more smoothly. I can't stress enough that everyone must take ownership in their position. This means not just being responsible for your part, but also realizing and appreciating the work others on the team are doing. This list of crew titles and responsibilities should help you understand how the team comes together and how each person individually contributes to the overall event.

Account Executive (AE)

The Account Executive is typically the person who has the relationship with the End Client, understands their vision, and manages the acquisition of the equipment and/or labor to execute an event. I like to call the AE "The Gatekeeper." They typically have all the information about the event and distribute it to the various departments, the Technical Director, and/or Project

Managers. Building a connection with the AE is key. Knowing the Account Executive means I have less of a gap, or filter, between the warehouse and the jobsite if there is an update, I need to make a change, or I need to add a piece of equipment.

Technical Director (TD)

The TD typically has relationships with the End Client, the Account Executive, and the production company. The TD often takes the vision of the End Client and Account Executive, then works with the Production Rigger, Lighting Designer, and the Audio Lead (A1) to turn that vision into a Vectorworks or AutoCAD drawing. The TD also helps to create the production schedule, then manages the on-site build of that vision. The TD is responsible for knowing what is and isn't available at the venue, and how that vision will work within the confines of the space. This is done through the building of relationships and organizing data collected from multiple parties. For a Video Projectionist, the TD is a key person because they oversee and connect with every department on the team.

Project Manager (PM)

The Project Manager works hand in hand with the Technical Director and often helps to secure any necessary equipment adds. If something I need isn't available on-site, the PM is my contact person. The PM is basically the on-site eyes and ears of the Account Executive. Most Account Executives have multiple shows going on at once, and a good PM helps everything to continue operating at a high level.

The Production Channel Podcast

On the Production Channel Podcast, there's an interview with Technical Director Roger Desmond. He talks about the importance of accurate and updated drawings, as well as being on top of every element on-site and at the venue. "You have to not just be looking at the step you're on, but the step three steps ahead. [You want to be] keeping that forward momentum and making sure you aren't going to hit a bottleneck because you have your head in the sand."

Engineer in Charge (EIC)

The Engineer in Charge oversees the Video Department and essentially works as a Project Manager for all things video. There are multiple sources that eventually pass through the image feeding a video projector from graphics, to cameras, to playback, and part of the EIC's job is to understand the signal flow and what cables and power are necessary to run the department. The EIC will also decide where the power distro will live and determines the best locations for the various people working in Video Village. On a smaller show, the Video Lead, or V1, serves as the Engineer in Charge.

Video Projectionist

The Video Projectionist is responsible for building the visual display aspect of an event using video projectors and screens/surfaces. The Video Projectionist will typically interact with every single person on the production team from Stagehand to Stage Manager because they are the last in line of a video source reaching the audience. The Video Projectionist will take the video signal/cable from the engineering rack and run it to the location(s) where the projector(s) will live. They will do the same with the power cable feeding the projector(s). The Video Projectionist will also color balance, or match, all the projectors to make sure the image is consistent with the source and machines feeding it.

Head Production Rigger

These critical people know and understand the infrastructure of the building/venue and are responsible for hanging the overhead components of the show. The Rigger's knowledge is what keeps everyone safe, and all of the equipment secure and stable. The Rigger knows where the pick points are, where the high steel is located, and what limitations the venue has in relation to supporting the structure that needs to be built. If there are elements within the EndClient's vision that won't work within the venue, the Rigger is responsible for talking to the Technical Director and making these issues known.

> If you aren't ready [for your part in the show] when they get to you, then you disrupt the flow of the day.
>
> Brent Armstrong | Owner, Alpine Rigging | Las Vegas, NV

Riggers, whether they are employed directly by the venue or part of the production company, have gone through safety training and must be certified to work on a set. Many hotels have an approved rigging vendor they work with because that person is very familiar with the property and its structural abilities. Riggers calculate weights, ceiling heights, and structural supports so the equipment can be flown correctly. It's essential that they work with everyone from Scenic to Audio, and be on-site, ready to do their part. Where the designers are the creative, right-brain side of the event, the Riggers are part of the analytical, left side.

Lighting Designer (LD)

The Lighting Designer creates the lighting palette, color tones, and movement of the fixtures. The LD brings the space to life like an artist painting a canvas, because without lighting, it's just a plain room with white walls. The best lighting draws people into a focus point and creates not just a show, but an experience. Underneath the LD is a Master Electrician (ME), who runs cables, hangs the fixtures, monitors power consumption, and helps to execute the LD's design.

The Lighting Designer often works with the Video Projectionist to ensure that the light and the lighting fixtures don't impact the projected images, or vice versa. For example, if the LD has installed a moving light fixture on the same truss as a video projector, the mover, as they are called, will cause the projector to sway, instead of keeping the image still and fixed on the screen/surface.

> **We aren't just there to make sure things go right, we are hired for the moments when things go wrong.**
>
> Richard Dunn | Lighting Designer | Canton, GA

These elements should be considered prior to any of us entering the space, so that we can adjust the drawings as needed. If we wait until we arrive on-site to visualize and understand this build, then we are already behind schedule. We know we must respect each department and what they bring to the production. As a team, we set our egos aside because we know that at the end of the day, it's all about the experience, the End Client, and delivering their message in the way it was meant to be conveyed.

Know the Different Roles

Richard Dunn, a Lighting Designer and guest on my podcast, talked about an event with Katy Perry where dozens of things could have gone wrong, but his team was able to overcome the challenges of the venue and the short timeline.

Video Assist (V2)/Projection Assist

These people serve as the right hand to their leads, similar to how a First Lieutenant serves a Lieutenant. Whenever I work a show, I like to identify the people who will be on my crew as early as possible so I know what to expect. Hard-working and responsible Assists can undoubtedly make a show better, and can make the job of a V1, EIC, or Video Projectionist so much easier. Primarily, the Video Assist works in the overall video department, and the Projection Assist works one-on-one with the Video Projectionist. Both positions are critical in managing the department's workload, thus impacting the event's success.

Audio Engineer (A1)

This is the primary audio technician, who works with the Production Company, Technical Director, and Account Executive to design the audio system for the venue. On-site, the A1 oversees the

Just a few of the many talented professionals that I have been lucky enough to interview for the Production Channel Podcast.

Top Row (L to R): Tucker McFall (photo credit: Michael Weinman), Robert Permenter (center), Richard Dunn

Bottom Row (L to R): Beth Forbes (photo credit: Anissa Williams), Drew Lawless installation of the audio, as they prepare to mix and maintain the right levels during the event. The way the sound feels in a space is an art that great A1s have mastered. The audience can sense when levels are off, it's too loud or too soft, or there's a slight ring in the system.

Listen to my interview with Steve Mitchell on The Production Channel Podcast for more information on this position. "A lot of what we do on a show is about managing risk, whether it's a machine failing, a projector failing, a microphone failing. We all know how bad it can go if you don't have something to back up [the equipment]," Mitchell said.

Audio Technician (A2)

These people assist the Audio Engineer with everything from installing the speaker clusters and monitoring the audio system, to pinning microphones on presenters before they walk on stage. Whether it's a full digital arena-style rack system or two wireless mics and an analog console, the A2 is the person the A1 depends on heavily during an event.

Video Switcher

The Video Switcher, who is often also the EIC or V1 depending on the size of the show, is the technician responsible for routing different video sources to the screens and calling up the image the audience sees. Often receiving cues from the Stage Manager, the Video Switcher helps to control the flow of the show as they sit behind the board and direct the event one, video input at a time. These inputs can consist of, but aren't limited to, cameras, graphics, or video playback.

In my podcast interview with Greg Hartung, you will get an inside perspective of this position and how to be successful in the role.

Camera Operator

A Camera Op is responsible for building and operating their camera to capture an event. The Camera Op uses their ability to tell a story by allowing the audience to see things happening that their naked eye may not see. This is done through shot composition and the anticipation of what's next. The Camera Op's image will often pass through the video switcher and be magnified on the screen of the Video Projectionist.

For a behind-the-scenes peek at what a Camera Operator does, listen to the podcast interview with Tucker McFall.

Stagehands

A Stagehand is a pivotal role on any production. They help with the load-in, the load-out, and float between departments during the actual event. A Stagehand's versatile nature allows them to help wherever necessary and to be an integral part of a show. Many people in this industry start out as a Stagehand, which is a great opportunity to learn how every department functions. The more you know about all the jobs involved in producing an event, the more valuable you are to the team. Although a Stagehand is one of the lowest positions on the crew, the great ones rise through the ranks and excel in the industry. Listen, learn, and be willing to work hard.

Graphics Operator

A Graphics Op builds, organizes, and operates the graphic elements for an event. Using programs like PowerPoint, Keynote, and sometimes Adobe After Effects, the Graphics Op works very closely with a Presenter to convey their message to an audience. A Graphics Op will also communicate with the Video Projectionist to make sure his image is properly aligned and color-matched with the slides/logos in the presentation.

Stage Manager

The Stage Manager is a supervisor position that typically sits at the Front of House, or Tech Booth, with the LD and A1. The Stage Manager organizes a show with a show flow, rundown, or script to ensure the entire crew executes the production in a fluid and cohesive way. By calling out cues, the Stage Manager sets the tone and pace by which the event happens based on the End Client's vision.

Know the Different Roles

There is a hierarchical way in which the crew is run, but at the end of the day, we are one team. No matter what position you hold, we are ultimately all in this together. We are working to make the best event possible and create an experience our End Clients will remember. When you take ownership of whatever role you are filling, you take pride in your work, understanding how it affects the greater good. Whether in management or a lower-tier position, that work ethic will be noticed by the other people on the team. For the Stagehands and Assists, that is the key to getting promoted and working your way up to the next position. For the TDs, A1s, EICs, Video Projectionists and Stage Managers, that is how you set the tone of your event, and how you help to influence the next generation of technicians coming behind you. My advice is to always do your best. Whatever the job may be, it's about projecting the best image possible.

Efficiency! It's a matter of efficiency. If you understand the chain of command, and who has the information you need, then you know who the right person is to ask the question you need answered.

I know the different roles on a project, and I know exactly who I need to address certain issues with. If we need to take our concerns up to the next level, we do that together. We efficiently collaborate and build.

A task once begun...DONE!

John Brewer | 5 Year Video Projectionist | Baltimore, MD

7 | Understanding Your Drawings

When you close your eyes, do you remember the location of your first show? I do. The Gaylord Palms in Orlando, FL was the location of my first gig in the Corporate Event sector of the Live Event Production Industry, and to this day, it's still one of my favorites. **The Osceola Ballroom, with its beautiful and vast layout, was overwhelming when I first entered it; however, over the years, I grew to appreciate the space.** Although I became comfortable within its walls, when I started working events where I needed to review drawings and notice all the small details, I once again entered unfamiliar territory.

Photo cred: https://www.marriott.com/en-us/hotels/mcogp-gaylord-palms-resort-and-convention-center/events/

I've worked in this venue on several occasions. I now know the space and the orientation of the room, which means when I'm looking at a 2-D drawing of the room, I know the layout and how it will need to be set. If I'm facing the stage at the end of the ballroom, the service hall is located on the house left side with several doors leading to it. I can visualize the Front of House, and I can also tell where the attendees will enter the space. I can see where the truss will be flown, where the downstage monitors will live, and where the backstage area will be. All the production spaces, like Video Village, Dimmer Beach, and the Greenrooms will be visible and clearly laid out. The drawing gives me an understanding of what already exists in the space and what will be built by the production team. But by visualizing the parts specified in the documents, I can turn that empty space into a reality in my mind.

What the Drawings Tell You

The "architectural" drawings you'll receive from the Technical Director are like blueprints of a home, with labels and symbols for entryways, wall placements, lighting fixtures, etc. They give a birds-eye view of the room in a two- or three-dimensional layout, almost as if you are looking down from somewhere above the chandeliers. If these kinds of renderings are new to you, you can increase your spatial awareness by studying virtually any type of blueprint (for homes, apartments, buildings) and comparing it to the real-life room.

I recommend Googling an image of a ballroom prior to working in it. This way, you can compare its original layout to how it will look with your gear and set pieces. The key is to familiarize yourself with the different parts of the room prior to arriving on-site. In this example, you can see the difference between an empty Osceola Ballroom and the drawings I received for the job.

Typically, drawings will come in one of three different formats: AutoCad (.dwg), VectorWorks (.vwx) or Portable Document Format (.pdf). Each file type must be opened by its specific program. The viewer version of these industry standard software programs is free and more than enough to analyze the drawings of a show. However, people will generally pay for the full version of the software if they will be designing shows or assisting with specific parts of the production layout.

Understanding Your Drawings

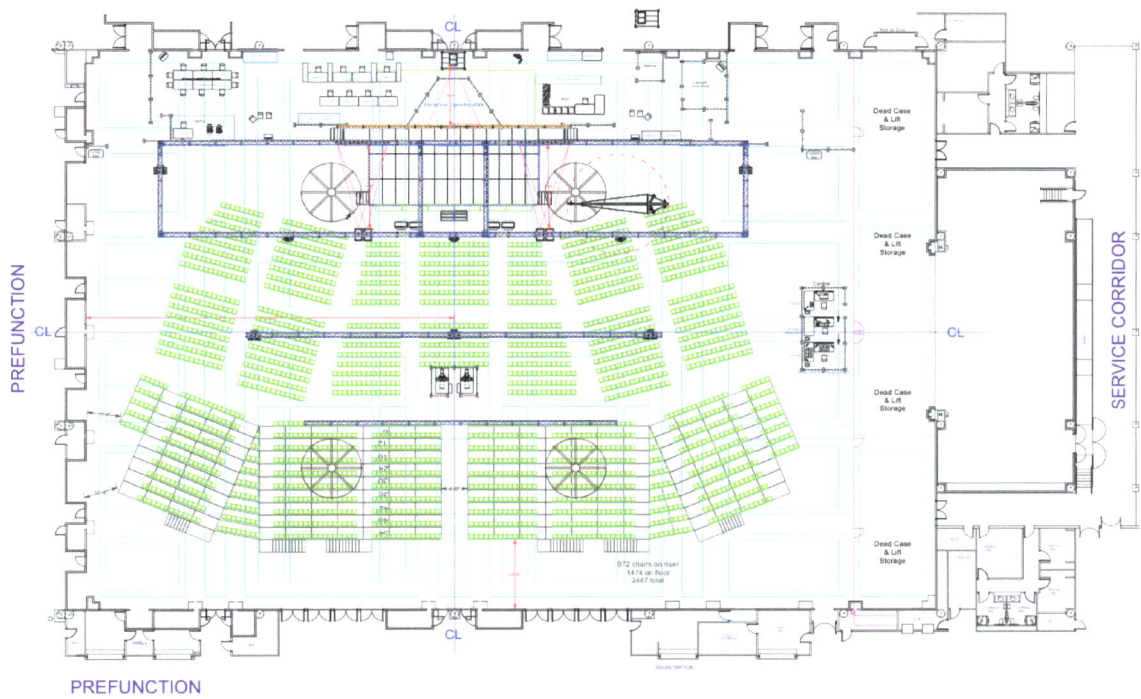

What You Need to Know

Once you have opened the file, there are a number of items you need to consider and begin to understand. As you go through this list of questions, think about your equipment list and where all the pieces will live in the ballroom. Even if you don't yet have all the details, you should always try to go into a job well-prepared and ready for anything:

Screens

- How many screens do I have?
  - → Knowing how many screens you have is step one in understanding your projector and cable needs.
  - → *This example shows 3 screens.*

- Are we setting up front or rear screens?
  - Will I have a dual vision surface on my equipment list? (these can be used for both front and rear projection)
  - *In this case, I have 2 front and 1 rear for my 3 screens.*
- What is the size of the screen?
  - This number will partially affect how far the projectors will live from the screen.
  - *In our example, the two front screens and the one rear screen are all 9'x16'*
- Will the screen surface be flat or curved? Are they wide-screen?
  - These factors affect whether you have to warp the image or not, and what software you will need to achieve your ideal image.
  - *This example has 3 flat surfaces.*
- Is the screen set within a scenic piece?
  - In the .vwx or .dwg file format, you can rotate the drawing to see different vantage points and possibly identify scenic pieces.
  - If you still can't tell from the drawing, ask the Technical Director for assistance.
- Are the screens ground supported?
  - What is the height from the ground to the bottom of the image?
  - Will the screen have a dress kit (draping) surrounding it?
- Are the screens flown from a truss?
  - What is the trim height of the truss?
  - What is the distance from the truss to the top of the image?
  - What is the distance from the ground/floor to the bottom of the image?
    - This will help you determine how high the screen will be above the audience's heads.
    - This will help you determine the height of the projectors.
  - *From this drawing at the Palms, I can see all 3 screens are flown from the truss.*

Projectors

- How many projectors are there total?
- Are the projectors single- or double-stacked?
  - → If double-stacked, are they stacked on top of each other?
  - → If double-stacked, are they stacked side by side?
  - → *In this drawing, the flown projectors are double-stacked and placed side by side, while the ground-supported projectors are stacked on top of one another. (Projectors are typically flown side by side to eliminate any line of sight issues for the audience.)*
- Are the projectors flown?
  - → Are they stacked on top of each other or are they side by side?
  - → What is the truss trim height?
  - → Are there moving lights on the truss? If so, I'll work with the LD to make sure there won't be any ballyhoo movements (to move quickly from one place to another) that will affect my projected image.
- Are the projectors ground-supported?
  - → Will it be a scaffolding or tower truss?
  - → What is the height from the ground to the projector?

A rule of thumb is if your projector is flown, you want it to live just inside the height of your image.

If the top of your screen is at 20 feet, then that is close to where you want your projection truss height to be as well. Something to always keep in mind, however, is there are unconventional seating arrangements.

For instance, are there bleachers in the room? An elevated seating space means you must make sure your projectors aren't in the line of sight for anyone in the room—especially, the person sitting the furthest away. Your drawings should have the height measurements and seating capacities of the bleachers.

- What is the throw distance from the projector to the screen?
    - → Using that same 9' x 16' screen in the earlier example, if I have lenses that have a 1.5-2.0:1 throw ratio, then my throw distance range is 24 feet to 32 feet. (16 x 1.5 and 16 x 2). The closest the projector can be is 24 feet from the screen, while the furthest is 32 feet.
    - → *On the drawing here, you can see that the throw distance is marked at 27 feet, and the rear projector is 19.5 feet from the screen. I need to make sure my rear projector lenses are close to a 1.1-1.5:1 throw ratio.*
- Where is Video Village in relation to the projector locations?

half the screen height + distance from the ground to the image = height of projector

Whether a scaffold or tower truss, I need to know the screen height in order to determine how high the projectors will need to be. There is a formula to calculate that information (above).

For instance, for a 9' x 16' screen with a 6-foot distance from the ground to the bottom of the image, you divide 9 in half, then add 6. (4.5 + 6 = 10.5) This means the projector needs to be placed 10.5 feet off the ground. The objective is to have the center of the projector near the center screen. If it's too low and pointing up, the image widens at the top like a V. If the image is too high and pointing down, it widens at the bottom

Cabling

- What will be the cable path?
    - → Identify the various paths from the projectors to Video Village
        - What are the best routes for cabling?
        - They should be unintrusive and efficient
    - → Confirm your vision of the cable paths with the TD's vision
    - → Are other departments using the same paths?
    - → What is the distance from the engineering rack/switcher to the projectors?
        - How much cabling will I need to reach the projectors?
        - Do I need to run multiple lines from location to location?

These are general cable path-related questions and should be thought out during your pre-production process. Another factor that should be considered is the location of your power drop.

- → Where is the house power drop?
    - Is the projection power drop coming from the same shared power distro as Video Village, or is it coming from another location like in the ceiling, which means it will live on the truss?
    - Are we sharing power with another department, like audio? Lighting often has its own power drop, because of the high load for dimming and effects.
- Where are the walkways?
    - → Where will the client/presenters and the rest of the crew members be walking back and forth?
        - How can we minimize obstacles?
        - Where can we install cable ramps and mats?
    - → Are there caterers and other people working the show?
        - Are there carts and hot boxes coming in and out?
        - Are there additional set pieces that will need to enter the space?
- What is the truss trim height?
    - Where will the breaks in the cables happen?
    - Do I need to add service loops?

What you begin to understand from looking at a drawing is that there is life happening and things moving within the 2-D space on the paper. In my mind, I can see the people working, I can feel the cases being pushed, I can hear the scissor lifts beeping, and I can react to the Rigger yelling, "Truss coming in!"

When thinking about the truss trim height, I understand I need to not

---

Envision all your cables stretched out for every projector: signal, power, and networking all run together side by side.

When you think about it, every projector has 3 cables, and that can add up to a lot of lines snaking across the room.

Remember, those are only *your* cables. What about the other departments? What about the cables for all of their equipment?

That can be a lot of loose cables if not organized and made neat. Therefore, it's important to think about the safety of others even before arriving on-site.

only measure the length of the cables running from the projector across the truss to Video Village, but also how far the cables drop to the ground. Not every cable will be a home run and be long enough to go directly from Point A to Point B. You will have breaks and connections throughout. The goal is to have as few breaks as possible, identify the ideal break locations, and label them on your drawing. This will help you minimize points of error and troubleshoot, should problems arise.

It is also important to think about service loops. When your trusses aren't all tied together like a grid, some sticks may float independently and need to be brought down by the Riggers. To do so, you will either need a cable break or a service loop the length of the truss trim height to the ground. This way, if the truss needs to be brought all the way "in," you won't have the possibility of snapping a cable.

As previously stated, by discovering the answers to these questions (or asking the TD for clarification so you can fully understand what you are working with), you are learning the show prior to arriving on-site. That is key to being proactive and productive in the eyes of your client. These are characteristics that represent a good brand and often lead to more job opportunities.

Downstage
The front of the stage, closest to the audience.

House Right/House Left
These are the sides of the stage from the audience's point of view looking at the presenters. An important note is that the House and Stage points of view oppose each other when you are referencing them in context.

Production Management Table
This space is designated as a working station for everyone who oversees the event. From the Producer, Stage Manager, Assistant Stage Managers, and Production Coordinators, this collective orchestrates the logistics needed to execute an exceptional show.

Projection Cone/No Walk
The light from the projector spreads out like a cone until it reaches the screen. If people walk through that light, their silhouette will be projected on the screen. To mitigate this interfering with the image of the presentation, this space is blocked off from foot traffic and used for low case storage.

Stage Left/Stage Right
These are the sides of the stage from the presenter's point of view as they look out at the audience.

Upstage
The rear of the stage, farthest from the audience.

Video Village
This is the backstage area where the video department sets up its equipment and controls all elements related to the visual presentation. When you locate this space and your projectors, you can plot out your cable path. You don't want to waste cabling and go to the house right side if the fastest, easiest, and least intrusive path is to the house left.

Other Components of the Drawings

You'll see other things marked on the drawings, such as the Make-up Area, the Client VIP Green Room, Furniture Storage, etc. You'll also notice Working Case and Dead Case areas are marked, showing where the still-in-use and empty equipment cases will be stored. The pink dotted line on the house right side of this drawing represents the drape that will hide the dead cases and scissor lifts when they aren't being used.

How to Annotate Your Drawings

Working with software like Vectorworks, AutoCad, or a PDF viewer gives you the ability to take measurements and add your own details to the drawings. For example, when I'm reviewing a document in Vectorworks Viewer, I can measure the cable path by clicking on the ruler, clicking on my starting point, then dragging the cursor from Point A to Point B. If I need to change the direction of my cable measurement, I can click the mouse again and continue to drag the curser along the new path. The numbers I receive will not only give me the total length of cable I need, but it will also help me understand where my cable breaks will take place. Then, I can check

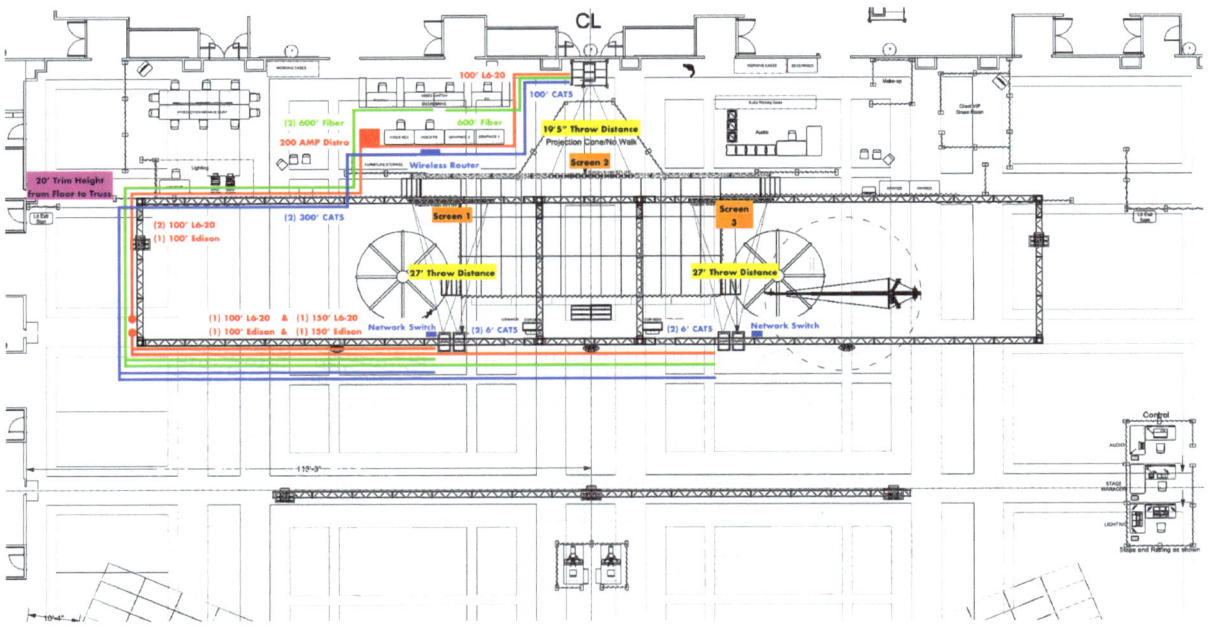

my measurements against what is supplied on the equipment list. By doing this prior to arriving on-site, I am setting myself up for success. I don't want to show up on load-in day and not have enough cable needed to execute my build.

After I collect all the details I need, I then use a PDF viewer to annotate my documents using various symbols and a color-coding method. I use green to signify video signals, red for power, and blue for networking cables. I also add all my cable breaks, service loops, trim heights, throw distances, and other pertinent information for the event.

This process that I use did not come together overnight. It was created after many years of doing the job, watching others doing the job, and wanting to figure out a way to do it better. As the well-known proverb goes, "Necessity is the mother of invention". It was necessary for me to scale my business and provide an infrastructure that others could learn from. It was also a necessity for me to pass on the knowledge that I obtained before I completely transitioned out of my role. Thus, the [Video Projection 101 Workbook](https://www.clemcoav.com/projection101/) (https://www.clemcoav.com/projection101/) was formed with the templates and documents to help you be successful and build confidence in your role.

Always Confirm and Check

Part of building that confidence is being secure in your vision and understanding as a Video Projectionist. I realized long ago that I can't complete my pre-production process without confirming with the Technical Director that their final versions have been drawn and uploaded for review. This is important because you want to make sure you aren't misinterpreting the drawings, working off an outdated drawing, or creating double work for yourself or anyone else.

Things *will* change, and that's why it's important to go into this process with a flexible plan. Whether it's the vision of the client, the available equipment, or the layout of the space, you must be ready for the unknowns. Start your planning process with the grace and understanding that things aren't always rigid and firm. You have to be fluid and adjust. You can't be so steadfast that you aren't able to adjust on the fly as needed. Nothing ever goes a hundred percent according to plan. That's one of the things that I truly believe has allowed me to be so successful—my belief that no one is perfect, and that other people are doing the best job they possibly can.

Understanding Your Drawings

Your mindset is far more powerful than you know. If things don't go as intended, just roll with it, because something even more beautiful can come as a result. That positive mindset can affect the whole room and the entire crew. Yes, we have a goal to be perfect, and yes, we want to create perfection. This is why we create a plan. However, just as in life, things rarely go as intended, and if you can have an onward-and-upward mindset no matter what, all of the people attending the event will benefit from that energy.

In our sample general session room, we have close to fifty people working the load-in, twenty-five opening the show, and 2447 people who are attending this conference. The way we, as crew members, interact and engage with one another will affect the entire experience for those 2500 attendees. If you are stressed or grumpy, people will feel it, see it, hear it, in everything we do from the lighting and how we are switching the show all the way down to the energy of the presenters on stage. Even as a Video Projectionist, your energy is transmitted through the screen and into the room. Think about it. Based on the quality of your work and the amount of effort and detail you put into it, people will either see the message clearly or contend with blurry imagery.

You are aligning and painting a picture in the ballroom, and the entire crew is collectively working to fill that empty space and create something amazing. Look at yourself the same way—how are you going to fill that empty space inside yourself? What will you offer to people who get close to you or work with you? It's within your power and control to decide whether that experience will affect others in a positive way or a negative one. After all, it's your show—understand your drawings and make it an impactful one.

> **The Drawing is the BIG picture of the overall concept. When the idea is sketched out, it makes it collective for everyone involved in the production to share and understand.**
>
> **When I first look at the Drawing, I want to understand the overall goal. Then, when I'm on-site, I share the Drawing with the Stagehands. If they understand the big picture, then it helps them complete their individual tasks. If you don't show them, then you could find them toiling away and doing things wrong.**
>
> **Understanding the big picture helps me to identify any potential problems and minimize having to do things twice.**
>
> Zamir Zeigler | 20 Year Video Projectionist | Hollywood, FL

8 | Understanding Your Equipment

Once you have your drawings and have researched the venue, it's time to begin reviewing your Equipment List. **Doing this prior to arriving on-site is paramount because your Equipment List tells you what gear and tools have been allocated for your job.** If you don't have everything you need and are trying to execute your build without the necessary resources, your product will suffer, and your client could be displeased with the end result.

This is why taking notes and making annotations are so important as you go through the drawings. Knowing how many projectors you should have, which lenses you'll need, the distance from the engineering rack/power distro to your projectors, and what types of cables you will need to make your cable runs are a few of the things you should understand from looking at your diagrams. Then with that information, compare your notes to the Equipment List and ask yourself, "Do I have everything I need to accomplish the job?"

There is a sample Equipment List from the Video Projection 101 Workbook (www.clemcoav.com/projection101) on the following page. Look at it. Get familiar with the layout, the terms used to describe the gear, and study how everything translates into an on-site gear requirement. Although Equipment List formats vary per company, the more you study the different types, the quicker you will be able to decode the information for your setup.

Understanding Your Equipment

Sample Equipment List

PROJECTION - 3 screens

3	Panasonic PT-RZ21KU Laser Projector (21K ANSI)
3	Lang - 20k Panasonic Rigging Frame (PT-DZ21K)
3	Lang - Panasonic Rigging Plate - Swivel Clamp (Gimbal)
3	Panasonic - Laser Projector Remote
3	XLR (F)-Mini 3.5mm (Projector Remote Adapter Cable) 1.5'
3	XLR (M)-Mini 3.5mm (Projector Remote Adapter Cable) 1.5'
3	3-Pin 20A/250V Power Cable (Panasonic 10K)
3	3-Pin Twist 20a/250v Cable (L6-20) 50'
6	3-Pin Twist 20a/250v Cable (L6-20) 100'
3	Case - Video - Projector / Panasonic PT-RZ21KU
3	Projector Rigging - Single Universal Hang Mount Kit (75lbs) (Panasonic RZ970, Sony VPL-FH500L)
3	Premier Mounts Low-Profile Single Universal Projector Mount (75lbs)
3	Mega-Claw w/ 1.5" Pipe Coupler (Black)
3	1 1/2" Pipe 1'
6	Lighting Safety Cable 30" Black 1/8" Aircraft Cable with Spring Hook
3	Case - Pelican 1550NF with Trekpak divider (Orange)
3	D75LE20 - 1.7 - 2.4:1 - Panasonic Zoom Projection Lens
3	600' Tactical Fiber DVI DA Package
3	DA - DVI - 1 x 4 - Kramer
6	DVI to Fiber Extender - Gefen FM 1000 - (SC / TX)
6	DVI to Fiber Extender - Gefen FM 1000 - (SC / RX)
3	600' 4 Fiber MM, Military Tactical Cable (SC-SC)
12	DVI-D Dual Link Cable / 6'
3	Cable Reel with drum extension
6	3 ft USB to Type H Barrel 5V DC Power Cable for FM1000
1	9 x 16 Stewart RP Screen Kit
	2. 9 x 16 Stewart RP Screen Kit

BACK-UP PROJECTION

3	Projector - 16:10 / Panasonic RZ21KU - HD - Laser (20K ANSI)
3	Panasonic PT-RZ21KU Laser Projector (21K ANSI)
3	Lang - 20k Panasonic Rigging Frame (PT-DZ21K)
3	Lang - Panasonic Rigging Plate - Swivel Clamp (Gimbal)
3	Panasonic - Laser Projector Remote
3	XLR (F)-Mini 3.5mm (Projector Remote Adapter Cable) 1.5'
3	XLR (M)-Mini 3.5mm (Projector Remote Adapter Cable) 1.5'
3	3-Pin 20A/250V Power Cable (Panasonic 10K)
3	3-Pin Twist 20a/250v Cable (L6-20) 50'
6	3-Pin Twist 20a/250v Cable (L6-20) 100'
3	Case - Video - Projector / Panasonic PT-RZ21KU
3	D75LE20 - 1.7 - 2.4:1 - Panasonic Zoom Projection Lens

From the departmental categories of gear to the quantities listed per item, your Equipment List is a source of truth to what will be available on-site.

Projection

The drawings will tell you how many projectors you have, and whether they are flown or ground supported with truss towers or scaffolding. The Equipment List will also tell you if the items have been reserved and if they are available for your event. For instance, if you have two double-stacked flown projectors on your drawing, you need to not only confirm you have four projectors, but you will also need to have four lenses, four rigging kits, etc. on your list. If two projectors are ground-supported, you should see the necessary scaffolding or truss towers on the list as well.

Are you familiar with the brand and model of the projectors? What is the maximum lumen output, or more simply, how bright will they be? What kind of power connectors do they have? Do you have the right cables for those connections? Lastly, what is the amperage draw? As you begin to develop your own process you will ask yourself similar questions to gain the required information.

Watts ÷ Volts = Amps

If the Amps aren't provided on the projector, you can use the Wattage and Voltage to calculate the number of Amps.

1800W ÷ 200V = 9Amps

1800W ÷ 240V = 7.5Amps

Lenses

How many lenses do you have for your projectors? What models are they? Are the lenses compatible with the projector model? What is the zoom range of the lenses? When you look at the drawings, you will know how far you need to project, or how far your throw distance will be.

A 9' x 16' screen is 16 feet wide. If your lens is 1.5-2.0:1, then your throw distance range is from 24 feet (16' x 1.5 = 24') to 32 feet (16' x 2.0 = 32').

Understanding Your Equipment

What you will then need to know is what zoom ratio is needed to fill the entire screen with your image. That distance between the projector and the screen will determine that.

If you have a 9' x 16' screen, and your lens has a 1.50-2.0:1 zoom, then multiply the width of the screen by 1.5 and 2.0 to get the zoom aspect ratio. With those results, will those lenses accommodate the throw distance you have between the projector and the screen?

Screens

What brand, what size, and how many screens do you have on your drawings? Now look to see what brand, what size, and how many screens you have on your Equipment List. What surface types are needed, are they front or rear projection, and what surface types are being supplied by the vendor? Does the Equipment List reflect what is shown on the drawings regarding dress kits for the screens? Or will the screens be fitted inside a set piece? If they are in a set piece or flown without one, does the Equipment List show the appropriate rigging supports to secure the screens to the truss: lighting safeties, Verlock/Griplocks, or shackles, etc.?

Rigging

As you may know, rigging isn't only used for hanging screens, but also for hanging projectors, running cables, and mounting other departments' equipment as well. That being said, some of the broader rigging items we, as Video Projectionists, should look for on the Equipment List are scaffold towers, truss towers, flown projector rigging, and screen supports, whether flown or on the ground. The more specific items include, but are not limited to, the correct number of scaffold uprights, crossbars, pigtails, screw jacks, sliding trays or wheels, sticks of truss, base and top plates, bolts, sandbags, cheeseborough clamps, Doughty double clamps, pipes, gimbals, span sets or steel safety cables, shackles, and tie line.

Reviewing the rigging portion of your Equipment List also means making sure you have the correct amount of hardware for the number of screens, projectors, and cable runs you will set up. If you arrive on-site and realize your numbers are off, you may be able to contact the in-house Riggers for additional resources.

When you are ready to hang your equipment and mount it to the truss, it is of the utmost importance to establish a working relationship with the on-site Production Rigger. They will help ensure all your gear is secured tightly all while making everyone's safety is a high priority.

Video Signal

The constant changes in the quality of video we are viewing also mean a change in the types of cables we are using to consume content. When I first started working in the industry, there were a lot more VGA, RGB, RGBHV/5 Wire, Composite, Component, and even S-Video devices. Now, we are seeing a lot more Fiber, HDSDI, SDI, HDMI, RF, and DVI connectors.

The quality of the video and the length of your run will dictate the cables you'll need to transfer the signal. If your signal is traveling a distance further than 300 feet, you will typically use fiberoptic cables because they allow for the highest bandwidth over long distances. Will you need converters to translate the signal style of the cable to that of the equipment? There are many types and brands of converters. Knowing what's considered a reliable brand and knowing you have enough primary and backup converters will make or break your show.

Power

Throughout my career, it seemed as though power was the most important subject to understand, yet the one thing people knew the least about. Seeking the knowledge for myself, I spoke with several technicians to try to pull together as much information as possible. Through my research, I found that the lighting department knew the most about this matter. I often reflect on conversations I had with Lighting Designer Heather Crowne and the numerous YouTube videos I watched trying to make sense of it all.

If your projectors pull 9 Amps each, and you have 6 projectors, then 18 Amps are required to run all of your projectors on 3-Phase power.

**9 Amps x 6 Projectors =
54 Total Amps (TA)**

**54 TA ÷ 3 Legs of Power =
18 Amps/Leg**

Understanding Your Equipment

What it boils down to is that lighting is mainly about amperage draw and balancing your load. This is why we identify the projectors we are using and looked to see what amperage they will draw, or pull, from the building. If you have 6 projectors, and each projector draws 9 Amps, then the total amperage draw is 54 Amps. When using 3-Phase Power, black, red, and blue are your hot legs, green is your ground, and white is your neutral. Divide your total amperage draw (54 Amps) by your number of hot legs (3 Amps) to make sure you are drawing 18 Amps per leg of power.

If you leave a 20% grace, or buffer, on your power drop to account for any possible surges or fluctuations, then a 60 Amp per leg power service is sufficient because you are only drawing 18 Amps per leg. This means you have enough power to account for the rest of your equipment, and the other equipment from the Video or Audio Departments. Similarly, with a 400-, 200-, or 100-Amp service drop, the same 20% buffer rule will apply. Your ultimate desire is to protect your equipment, your show, and prevent any moments where breakers are tripped because of spikes in your power draw.

Now, after establishing the amperage service drop needed to operate your projectors, check your Equipment List to confirm you have enough cables, ports on your Power Distro, and the correct connector types for your setup. It's important to identify the connectors used on your projectors and distros to prevent any issues during your install. Some of these connector types include, L-2120, L6-20, L6-30, and Edison. If you have 6 projectors with L6-20 connectors, then your Power Distro should have at least 6 ports for those same types of connectors. If that's not the case, then you can either replace the Distro, order a second one, or potentially share power with the Audio department.

Cables

One of the things I think about the most when comparing my Drawings to my Equipment List is efficiencies. A way to simplify your setup is to know what cable lengths you need and draw out your paths during your pre-production process. Whether it's figuring out your feeder cable gauge and path, running L-2120 with Edison breakout boxes, or choosing to use Socapex as a multicable option over individual cable runs, finding the best way to power your equipment is the goal.

I prefer to use a Socapex, which is a 19-pin electrical connector, over individual cables because it allows me to run one cable instead of six. Though it is significantly heavier, Socapex has fan-in and fan-out adapters that provide the option for six Edison or six L6-20 connectors to be plugged into the distro and then run to power the projectors.

Though these questions and details may seem small and tedious, every single one adds up. Every single detail ensures you can do your job, and every single detail ensures you can do it well. The last thing you want is to be on a show site or out of town, and not have the tools or equipment you need. When you don't prepare and do your pre-production work, you're setting yourself up for failure, and you're setting your client up for a difficult show.

If you need help locating and identifying information as you prepare for your upcoming event, use your network and resources to acquire what you don't have. Visit our VPSG Facebook Group if you have questions or need additional support. There are a number of professionals there who would be happy to answer your questions. For a deeper dive into show power, I highly recommend *Electricity for the Entertainment Electricians & Technician* by Richard Cadena.

I have always focused on this because there is no room for error. You must know your Equipment List—PERIOD!

Your EQL gives you the general knowledge about the tools you have to be prepared and do your job. Your power, your lensing, your signal source...all of this is there.

You don't want to be on a job where you're forced to run HDMI cables to your projectors because you ran out of SDI. I did it, and it wasn't fun.

Kevin Rose | 28 Year Video Projectionist | Las Vegas, NV

9 | Understanding the Production Schedule

There are three critical components to your pre-production process, two of which we have already discussed: the drawings and the equipment list. The third, **the Production Schedule, is equally as important as the previously mentioned documents.**

The Production Schedule is a timeline for every department, every crew member, and every talent/presenter working the event. It is a massive undertaking to organize the number of people and elements on a production, and sticking to a timeline is crucial for seamless execution. The Production Manager and the Technical Director work together to compile the Production Schedule and then distribute it to the various crew members.

I typically print the Production Schedule when I first receive it, put it inside my Projection 101 Workbook, highlight it, and make notes next to key areas. I then review the schedule on my travel day, leaving me plenty of time to get acclimated to the hotel or venue and get a good night's rest. Working a show may mean many long days and working late into the night. Starting a job well-rested can help maintain a productive energy level throughout the event.

Gaylord Palms - Orlando, FL   MASTER PRODUCTION SCHEDULE v1   9:37 AM, 1/10/22
                                          OSCEOLA BALLROOM

WEDNESDAY, MARCH 11		NOTES
5:00 AM	**CREW CALL: TD**	Osceola 1-6 & AB
5:00 AM	TD Tom Bollard reviews rigging plans & pre-marks floor	
6:15 AM	C.AV Lighting Truck #1 at dock	
6:15 AM	Catering: beverages for crew/replenish thru-out day (7a - 11p)	Place in room per Tom Bollard
6:30 AM	**CREW CALL: In-house AV Rigging/Lighting (2 lead + 5 hds)/Loaders (6)**	Osceola 1-6 & ABCD
	In-house AV begin rigging & bolting truss	
	Electrical install motor power	
	C.AV unload lighting gear and start moving to ballroom	
7:00 AM	**CREW CALL: Lighting, Audio & Video Leads**	See Call Sheet
	Meet with TD to discuss plans for install	
7:45 AM	C.AV Audio / Video trucks #2 & 3 arrive at dock	
8:00 AM	**CREW CALL: Lighting, Audio / Video (all others)**	See Call Sheet
8:00 AM	Electrician to meet TD to install power drops where possible	
8:45 AM	Scenic Truck arrives at dock	
9:00 AM	**CREW CALL: Scenic (2 leads / 10 hands)**	Osceola 1-6 & ABCD
11:00 AM	End of Day - Truck Unloaders	
By Noon Goals:	All trucks unloaded w/ gear in room, truss at working height, C.AV gear attached to truss in all areas, cabling started	
11:30 AM - 12:00 PM	Catering: Crew Meal Break: Riggers & Lighting (30 mins. on clock, provided)	Osceola 1
12:00 PM 12:30 PM	Catering: Crew Meal Break: Audio & Video (30 mins. on clock, provided)	Osceola 1
noon-4:30pm	Continue in all departments as needed to hit daily goals	
By 2:00 PM Goals:	All truss over stage area flown to safe working height for stage build	
2:00 PM	Stage truck at the dock with their own labor	
3-5:00 PM	Stage built and carpeted	
4:30 PM	Catering: Crew Meal Break: All crew (30 mins. on clock, provided)	Osceola 1
5:00 PM	Drape Company truck at dock with their own labor	
5-9:00 PM	All drape installed	
9:00 PM	EOD for all crew except audience riser team.	
By 9:30 PM Goals:	Backstage buildout done w/ engineering functional & faxed	
	All lighting, audio and video gear installed onto truss and cabled	
	Truss at trim	
	Stage built and carpeted	
	Drape install complete as available	
9:00 PM	Security Begins Overnight (1 person)	
9:00 PM	Audience riser truck arrives at dock w/ their own labor	2 forklifts
9:00 PM to 6:00 AM	Audience riser installed and completed	
4-7:00 AM	House install chairs	See TD floor plan

THURSDAY, MARCH 12		NOTES
7:00 AM	IT tech to meet TD to finish install of internet drops where needed	
7:45 AM	Catering - beverages for crew/replenish thru-out dy (8a - 12a)	
8:00 AM	**CREW CALL: GS Crew**	See Call Sheet
8:00 AM-4:00 PM	Complete work in all areas (focus/ programe lighting, align projectors, tweak audio, overall cleanup of all areas, tape stage, etc.)	
9:45 AM	Furniture Company truck at dock	Del to Osceola 1
11:30 AM	Crew Meal Break: Scenic (30 mins. on clock, provided)	TBD
12:30 PM	Crew Meal Break: ALL (30 mins. on clock, provided)	TBD
1:00 PM	In-House Engineer to turn off lighting breakers as needed, see LD	

Day One

This is the first day of load-in, and if you look at the sample schedule, you'll see that the TD has a 5:00 a.m. call time. The rigging and lighting departments start at 6:30 a.m., followed by the video, audio, and remaining lighting crew at 7:00 and 8:00 a.m. As people and trucks continue to arrive, there are multiple discussions about the day's expectations, and the equipment is consistently offloaded.

Out of the many venues I've worked in, I know this particular ballroom and hotel very well. I know where the service elevators are located, how long it takes to unload the trucks and get the equipment to the elevators, and then from the elevators to the ballroom floor. From there, I can look at the schedule and see that we have a realistic timeline for these tasks. If you haven't previously worked in a venue, I suggest you do a walk through the day before. This way you can familiarize yourself with the path the equipment will take, then get a rough estimate of timing. Remember, the trucks are bringing equipment for all the departments. There will be a fair amount of chaos in the unloading process. You will have to be quick on your feet, attentive, ready to work, and prepared with a checklist to make sure everything you need has arrived.

When you do a quick glance at the Production Schedule, keep in mind your meal breaks. Something as simple as lunch can make a difference in your schedule, budget, and overall demeanor. Is it catered on-site, or are you on your own? Do you have a half hour on the clock or an hour off the clock to eat? This particular client has all-day beverages and staggered thirty-minute meal breaks. This allows the work to continue, encourages the crew to focus on the project, and keeps the day moving forward with minimal distractions.

Studying and analyzing your Production Schedule, even at home, will make your upcoming on-site project that much easier.

Something to note on the schedule are the KPIs (Key Performance Indicators). What expectations do the TD and client have of me, according to the Production Schedule? What do I need to get done, and what is the timetable for those tasks? On the sample schedule, all trucks have to be unloaded, equipment transported to the ballroom, and the trusses at working height by noon.

Working height for the trusses means they are bolted together, attached to motors, and are floating three to four feet off the ground. This allows the various departments to lay cables and attach their equipment in the designated locations according to the drawings. I know I have four projectors that will be flown on trusses. Therefore, I need to get with my crew to ensure we are getting our work done and not holding up the entire process. There is a domino effect if a department gets delayed, and that can drastically change the workflow.

This is where the prework becomes important. If you have already studied the Drawings and Equipment List, then you know if you have enough cabling to get from point A to point B and how best to run those lines.

According to the schedule, we have until 2:00 p.m. to make sure the main truss grid is flown and out of the way. The carpenters will be on-site building the stage from 3:00 p.m. to 5:00 p.m., and as a result, I need to have my screens assembled, in place, and attached to the truss prior to the motors moving. This is because all three screens are upstage of the deck.

The intent is for the stage to be completed before dinner so the drape installation can run from 5:00 p.m. to 9:00 p.m. After such, and once the crew has left for the day, the audience riser team will begin installing the raised seating. The schedule will continue overnight without any wasted time and go nonstop, because there is so much to be done before the first walk-in.

Day Two

Crew call is at at 8:00 a.m., and the load-in will continue until 4:00 p.m. The EIC and other video crew members will test all equipment, and the projection team will complete the grid alignment and color balancing. Our goal is to troubleshoot any issues and to be ready for the rehearsals that start later that day.

When the Stage Manager arrives for their 4:00 p.m. review, I know I will switch to my in-show mode. This is when we begin our technical rehearsals prior to talent and presenter arrivals at 5:00 p.m.

As I look over the Production Schedule, I see these two load-in days will be long but productive. Day One is sixteen nonstop hours for some, and Day Two is fourteen hours for others. It is critical to get a good amount of sleep every night, especially on Day Two, so that you are rested and ready for the first day of show.

Show Day begins with a 6:00 a.m. call for the entire crew, and another round of rehearsals start at 8:00 a.m. before the doors open. Meals are also staggered so the talent can sneak in an extra rehearsal if needed, and any problems can be addressed.

This is another long day, and one where you can't drag your feet. You must be attentive and aware of what's going on at the Front of House, in the Audience, and Backstage so you are available to assist wherever you're needed.

Remember, you were hired to work the show for those particular days, and that means you are available even outside of the regularly scheduled hours. I've been called in or had to stay on the clock way past what was originally scheduled. I am not resentful or resistant because that's my job—to be there when they need me and for as many hours as necessary. Remember, just because the Production Schedule says the day ends at 7:00 p.m. doesn't mean it will always be the case. Fortunately, overtime and double time will be paid as it is earned. The key is to be flexible and eager to help. A good attitude will go a long way in getting you recommended for the next job.

Load-Out

After the show is over, everything is pretty much done in reverse for the strike (dismantling) of the equipment. You're undoubtedly tired by now, but this isn't the time to rush. Make sure you're being thorough so you don't lose anything and you are being patient with the other members of the crew and staff. Throughout the load-out, you want to be safe and cautious so that no one gets hurt and no equipment is damaged.

When Things Go Wrong

Anticipate that something will always go wrong. Whether it's one department running behind and bumping up everyone else's schedule or a broken piece of equipment, you have to be prepared for anything. That's where those relationships with the rest of the crew can be important. A good relationship will allow for better communication and a more efficient installation process. Most of all, realize that at the end of the day, we're all just trying to do our best.

In 2013, I worked a show in Phoenix, AZ with two 100' curved screens. When we began warping the image coming out of the projectors, they started to glitch because of a software update. After troubleshooting the issue over and over again, we finally decided to replace all twelve projectors with an older model to get them to stabilize. Those fifty-five hours worked in three days weren't built in the schedule, but a solution had to be found. I went into that show, as I do every show, expecting something to go wrong, just not that wrong. By doing so, I was already mentally prepared to work overtime and expect long days. I always remind myself that "it is my financial pleasure to be here."

When there is a problem, the first step is to communicate. Whether you notice an unrealistic timeline in the Production Schedule or an issue arises on-site, go to the Technical Director and express your concerns. If the problem is within your department, talk to your Lead. Once you identify the issue, work together to come up with a solution and then act on the decision. When you are dealing with dozens of crew members, a multi-million dollar facility, and hundreds of moving pieces, there isn't enough time to sit around and wait for a solution to come to you. You must actively pursue it.

Sometimes working a show means long days and short nights, and you must learn to pace yourself. Take the breaks and get the rest you need that will help you endure for the run of show. Remember, when you are on-site, you are being asked to give your all to the production. You can't do that if you don't prepare yourself for those grueling days. At the end of the event and when you're done with the show, that's your time to go home and rest. It's important to build yourself back up to go do it all over again.

> **I remember a security officer at the Gaylord National once told me, 'You have to be like a garden hose—flexible. If you're not, you'll burst.'**
>
> **Production Schedules are a starting point and always changing. You can look at it and study it, but don't look at it as rigid and fixed. Your 6pm end-of-day could easily become a 10pm out.**
>
> **My advice: don't get frustrated and upset. That won't help anything. You have to be patient so that you don't become one.**
>
> Mike Swinton | 30 Year Video Projectionist | Orlando, FL

10 Teaching Others to Project

Once you have mastered the techniques of Video Projection, you should start giving back to the community that helped you get where you are today. **Many of my colleagues and I take pride in being able to share our skills, knowledge, and understanding with people who are just entering the industry**. This is because we know we wouldn't be where we are without the help of so many others along the way.

Teaching someone how to project is an art, just as video projection itself is an art. Yes, it's systems, steps, and processes, just as we have outlined in this book, but it's also the artistic understanding of what you are trying to achieve. You have to be able to look at every show or event with a creative eye, and to truly see the vision of the client.

As a Video Projectionist, you are taking information that's been packaged together about the screens, the projectors, the equipment, the venue, and the show itself, and bringing it to life. By having a thorough understanding of all parts of the process, you can get from A to Z with very few complications.

However, the key is to understand no one does it alone. Great productions are a team effort, created in a team environment. Just as the scaffolding and motors are there to support the trusses, projectors, and screens, so too are the people you work with. Everyone—Video, Audio, Rigging,

Teaching Others to Project

Lighting and Scenic—is there to be the scaffolding, if you will, for the event. We are all there to support each other, and it's only by working together that we can pull off these great events.

Set Up for Success

I've been in this industry for a long time, and I know the clients, the jobs, the venues, and the equipment so well that I can recognize cases as they are unloaded from the truck. Whether the shop labels them or not, you will eventually learn what to look for and do a mental checklist as things arrive. As you develop this skill, you will see the benefit of the pre-production process and quickly improvise when necessary.

With experience, you learn the systems that work for you. For example, I separate my cases into groups. All of my projectors are in one area with the lenses sitting on top. All of my cable cases are in another area close by, and all of the rigging isn't too far away. One of the companies I often work with writes the lumens of each projector on a quality control document inside of each case. This makes it easier for me to figure out how to average the light across the room and ensure an evenly balanced image. I then use my Video Projection 101 Workbook to organize the lenses I need, what zoom length they need to be, and which projector they will go with. I also have a list of the cables I will need with the appropriate labels already printed out. This is possible because

This picture was taken in 2016 when I first had the desire to teach and build my curriculum. Located at the Marriott Wardman Park in Washington, D.C. This TEK Productions event is where I first met my good friend, John Brewer.

I've done my pre-work with the desire to be efficient and know what cables go with which projector for that specific cable run.

I am doing all this preliminary work to set myself and my team up for success. The more preparation I do on the front end, the easier the day is going to be. I don't want anyone, including myself, to have to do double work or to go back and fix a mistake. My job is to gather and provide the information everyone needs to be successful. This way, we can project the best possible image for my company and especially for the client.

When I work with others, I assess their strengths and assign people to tasks based on their organizational or technical skills. I'm working right there alongside them, laying cables, building the screens, and rigging the projectors to construct the event. I'm the leader, yes, but I'm also not afraid to get in there and do the work myself. At times, I will step away and empower my team to continue without me, but we are a unit. From the minute we begin to unload the truck to the second the last case is loaded back on, we are one!

Having the Right Mindset

One of the things I firmly believe is that no one person is higher than another. We are all equal as humans, working on this project together. Some might have more of a knowledge base than others, but we are all equal.

Having that mindset allows us to work effectively across departments. No Lighting Director, Engineer in Charge, Technical Director, etc. is greater than a Stagehand or Camera Operator. If

"Your image is built on how you interact with people, how you do your work, and how prepared you are in good and bad situations."

Clem Harrod | Owner, Chief Projection Officer, CLEMCO.AV | 30 Years Industry Experience we all go into the project with that mindset, we will complete the job better and more effectively. The End Client will be happy, which means we'll all have an opportunity to work together again. It's in everyone's best interest to have a positive, can-do mindset.

Know What Works

As you are teaching your Video Assist or a brand-new Video Projectionist how to do the job, be sure to give them more than just a rundown of the equipment or the layout of the ballroom. Being a great Video Projectionist is about taking the necessary time to check and double-check everything, especially from a safety standpoint. These are some things every Video Projectionist should look for, or be aware of, during their install:

- Before firing up your projectors, you should meter your power cables to ensure you have clean power and none of your wires are crossed. An electrical disaster can happen if you don't.
- Factory reset your projectors to get a clean start for your setup.
- Test the network and signal of your projectors when they are still on the ground, not at trim height, to make sure the cables are properly connected and working. The last thing you want to do is re-run cables when the projectors are twenty feet in the air instead of five feet off the ground.
- Make sure the screen is level and flat. Trying to manipulate a flat image on a warped screen isn't fun.
- Place your projectors, whether front or rear projection, in the center of your screen. It is far easier to align an image when it's hitting the screen squarely.
- Once your light source is passing through the projector and you can see the shape of the projection cone, make sure it's hitting the screen correctly with a grid test pattern. Do you have a rectangle with four ninety-degree corners, or is it more like a trapezoid or parallelogram?

When you have 10,000 hours of experience (which Malcom Gladwell theorized in his book *Outliers*), you can begin to manipulate the projector in your head before you even touch it with your hands. This is because you have the experience to know exactly what to do to square up the image prior to actually doing it. Those instincts will tell you a trapezoid shape could mean the projector is too low in the front and too high in the back, or that a parallelogram means something completely different. Yes, you can go inside your computer program and warp the image and light source, but in the end, you aren't creating an image with the utmost integrity and producing the best possible product.

Integrity is very important in this industry. Your word is your bond. If you book a job, start a project, or make a promise to a client, keep your word. Your image is built on how you interact with people, how you do your work, and how prepared you are in good and bad situations. The client is trusting you with their image, and you must treat it as importantly as you do your own.

This is a small but growing industry, and those who aren't too proud to ask for help, understanding, and guidance will do well. Look at those above, around, and coming up behind you as part of a bigger team, and you will treat them with fairness and grace. In short, if you want to project an image of integrity, you need to live a life of integrity.

See it... Believe it... Achieve it! #Projection101

In my day, we were always so competitive. It was about what's the next big show, and who's on it. This was especially true when digital projection became a thing.

I can remember a specific Projectionist who ran a tight ship and made me write stuff down. I wouldn't say there was a true structure to teach me a process, but information was definitely passed along.

When it came to working with you, Clemmy, it was more about a rapport and relationship. I knew I could trust you, and I knew in my absence you would get things done. We didn't need to manage egos.

I knew if we worked well together, we would get to those big shows with the cool new projectors—TOGETHER!

That's what I'd say about *Teaching Others to Project*. Understand you can't do it on your own. If you want to get better and do larger projects, you have to teach others how to help you, and at the same time respect what they already know.

Phil Licari | 35 Year Video Projectionist | Orlando, FL

Thank you!

Thank you for taking the time to read, listen to, and embrace my inner thoughts. It takes a lot to share yourself with others without fear of judgment. Or better yet, to understand that the judgement that comes from man does not equal the judgment of the Father for not doing as you're told. Today, I have done as I was lead, or called, to do.

Back in August of 2019 I approached my career as an author to write what was to be my first book, the *Video Projection 101* Textbook, as a resource to support my Video Projection 101 Workbook. However, God had a different plan. Instead, my attention and focus went towards writing *Career Projection 101: An Independent Contractor's Guide to a Successful Business and Balanced Life*. I didn't know why, and I didn't understand how I could be pulled away from something that was clearly my gift. But, I walked in faith and began my journey.

There were obstacles and barriers that I had to overcome, and in the end a huge part of my journey was completed and documented. When I say documented, it's more like journaled. My story was told and written in a way that would help educate others on how to navigate their own journey, career, and life based on my successes and failures.

Being an Independent Contractor, or Small Business Owner, isn't easy. It is difficult—VERY difficult. Living in a gig-based economy isn't easy either. It too is very difficult. But with the right support and a strong sense of faith, anything is possible.

I presented my gift to the world, and my industry, on April 3rd, 2020, right at the start of the COVID-19 Pandemic. A pandemic that shut the Live Event Production Industry, my industry, down. All mass gatherings and live events were brought to a sudden halt and event after event was being canceled. Those cancellations would drastically affect people's livelihoods and incomes.

What I didn't realize at the time was that a pause and a reset was needed. We needed to step back and re-evaluate how we were living life. We needed to step back and see how certain aspects of our lives were suffering trying to live up to the demands of other people and companies. We needed to step back and realign our PROJECTED IMAGE.

What is YOUR projected image?

Where are you going?

Who will you become?

How will you get there?

When will you reach your final destination?

And...

Why are you looking to reach this place of Eden and inner peace?

These are all questions we have to ask ourselves and questions we have to teach and encourage others to ask as well.

This life is a journey, and we must understand how to recover from our Fall.

Will you take the time?

Will you take the time to understand your life before you've finished living it?

Will you take the time to understand how to align and adjust your projected image?

I know I will. See it... Believe it... Achieve it! #Projection101

Be well,

One thing I have learned through my thirty years of experience is that the gear is just the gear, but the people operating it make all the difference. If you treat them well, they will take care of both you and your client.

-Clem Harrod

Clem Harrod serves as the owner and Chief Projection Officer of CLEMCO.AV, which he established in April of 2016. Clem began studying the art of Television and Event Production in middle school in the early 1990s. After graduating high school, he began working towards his Bachelor of Science in Media Production at Florida State University and was an active member in the College of Communication's professional production group known as Seminole Productions. It was there he learned three key skills: the art of storytelling and seeing things for more than what they appear to be on the surface, the ability to anticipate and predict outcomes through shooting sporting events, and how to stay focused and attentive while working in energetic and entertaining environments.

Upon his graduation from Florida State in 2001, Clem began an internship with the Orlando Magic. There, he established himself as a very talented Sports Videographer and continued shooting NBA, MLB, NHL, NFL, and various other NCAA events for fifteen years. Through his contacts in the broadcast industry, Clem was introduced to, and simultaneously ran a career in, Corporate Event Production. Here, he began as a Stagehand and, through his strong work ethic and friendly personality, quickly worked his way up the ranks to become a very skilled Video Projectionist. With a new career focus, Clem decided to retire from the sports broadcast sector of the industry and devote his time to teaching and growing his own business. Now, CLEMCO.AV partners with many well-known production companies and their Fortune 500 clients to produce events for audiences of up to 15,000 attendees.

After leaving sports, Clem's desire was to understand how to best manage the personal and professional side of the Freelance/Independent Contractor lifestyle. Enlisting the help of various service providers, Clem created a platform of communication and networking that offers a solution that didn't exist when he started his career. With the newly found sense of balance and work life integration, Clem is now able to spend quality time with his family and educate others on the philosophy known as Projection 101.

Clem is very passionate about, and has a strong desire to share, his knowledge and understanding of Life Projection and Independent Contractor/Small Business Owner Lifestyle Management. If you would like to arrange for Clem to speak at your next event or have him work with your technicians and teach his methods on Video Projection, contact us at info@clemco.net.

Photo by Rich Johnson

About CLEMCO.AV

CLEMCO.AV is a labor consulting and networking resource for Independent Contractors in the Live Event Production Industry. With mixtures of career guidance and individual brand management to payroll, tax, financial and insurance providing services, CLEMCO.AV is here to assist you in your success. By gathering this information and resource professionals, we want to be a one-stop shop for Independent Contractors needing help.

In addition to helping Independent Contractors find the tools they need to have a prosperous career, CLEMCO.AV prides itself on being a collective of highly skilled, motivated, and client-friendly Audio-Visual Technicians. These Certified AV Techs have experience in a variety of in-show environments, can be trusted to put forth their best efforts, and will ensure your show's success. We will work with you to make sure your clients, and their End Clients, are happy.

"Our team understands that not everyone is passionate about what they do. Everyone doesn't feel the connection to the bigger picture, nor do they understand how their small piece plays a role in the project's overall success. At CLEMCO.AV, we do. Allow us to help you project the best image possible."

Abut CLEMCO.HR

CLEMCO.HR is a platform used by Independent Contractors, Freelancers, Entrepreneurs, Small Business Owners and individuals looking for Human Resource-style support. Our team of vetted service providers are standing by to assist in the areas of:

BUDGETING & BOOKKEEPING | ACCOUNTING | FINANCIAL PLANNING | HEALTH INSURANCE | BUSINESS INSURANCE | MENTAL & PHYSICAL HEALTH

If you are looking for assistance, visit us at www.CLEMCOHR.net

Why I joined CLEMCO.HR

As a health insurance advisor, I support independent contractors and families by researching all health insurance options nationwide. I joined CLEMCO.HR as a service provider because this platform directly connects me with people seeking advice and help. By specializing in self-employed contractors, entrepreneurs and their families, I can offer affordable rates to people who do not have access to employer-sponsored health insurance.

I genuinely care about people and serving them in the best way possible. CLEMCO.HR allows me to do that and creates a personalized human resources department experience for its users. CLEMCO.HR connects you with a team of trusted and vetted service professionals and having conversations with them saves you a considerable amount of time. In addition, CLEMCO.HR makes sure its users have all the benefits a W2 employee has at any large company. The service professionals will educate you on the plans and details that relate to your situation, then they will customize a solution for your needs as an independent contractor. CLEMCO.HR bridges the gap of what an employer does for employees by fostering an environment where you can do it for yourself. The service providers of CLEMCO.HR are meant to serve you throughout your career and over your lifetime.

Jessica Barnhill, CLEMCO.HR Health Insurance Advisor

How CLEMCO.HR has helped me

Clem Harrod, owner of CLEMCO.HR, has taken the time to listen to my dreams and goals. From there, he's helped me shape them in a way that makes them easier to achieve. Clem has motivated me to push past self-limiting beliefs and get over frustrations that I had with myself and others. He's also changed the way I view the world, by telling me to enjoy the process and focus on the little Ws along the way.

Clem says, "Every day is filled with small battles that we have the power to overcome, but sometimes we fall short. It's human to fall short, and that's why we have to focus on the small wins throughout the day. When you take those small Ws and add them all up, they equal a big W and overall success."

Creatives, like myself, often struggle with having too many ideas but not enough action. This is a result of a lack of structure and support to see an idea through. This group of tax experts, insurance advisors, mental health specialists, and financial planners will help you plan your career, save for a bright future, and achieve the business goals you've been losing sleep over.

Since joining CLEMCO.HR, I have opened a college fund for my daughter, finally have health insurance, and started saving for retirement. Setting these up was stress-free and maintaining them is easy because it's automatically taken out of my bank account. This was very important to me, as a new parent, because I know that saving early and often is vital when it comes to college and retirement.

I encourage all of my fellow freelancers and entrepreneurs to use CLEMCO.HR as a go-to independent contractor resource for all tax questions, insurance needs, mental health concerns, and financial planning. You will be happy you did.

Kelly Foxen, Foxen Productions

About CLEMCO.U

"When I speak to others about my journey, I try to always be open and honest as I share the lessons I have learned along the way. These stories are also being told in my bodies of work, both on the screen and in print.

My number-one goal is to help people gain an understanding about themselves, their projection, their entrepreneurship journey, working with others, and growing into their purpose. That's CLEMCO.U—Coach, Lead, Educate, Mentor. This platform, and these engagements, are the means by which I share my understanding.

I truly believe students, recent graduates, entrepreneurs, and people from across all industries can benefit from hearing the journey of how I have learned, and continue to learn, to navigate a career and a life that I love.

It's not just about doing the work—it's about being the best version of yourself, in all areas of your life. It's about maintaining the balance of work and family.

It's about Seeing it...Believing it...and Achieving it!

It's about Projection 101."

For more information on booking Clem Harrod as a speaker at your next event, please email guest_speaker@clemco.net.

FOLLOW US ON SOCIAL
@CLEMCO.AV

Professional Projectionist Clem Harrod
brings over 30 years of knowledge and experience to his

Coach, Lead, Educate, Mentor

platform where he focuses on helping others in the Live Event Production Industry
enjoy a career with less stress and more financial opportunities

JOIN OUR FACEBOOK GROUP!

ABOUT OUR FACEBOOK GROUP

Video Projectionist Support Group (VPSG) is a co-op for Video Projectionist who interactively collaborate as a valuable resource of technical support. From power and signal, to lensing and networking, and everything in between, VPSG is a closed community where we come together, stay connected, support one another, and grow in the industry.

www.ingramcontent.com/pod-product-comp iance
Lightning Source LLC
Chambersburg PA
CBHW041102070526
44583CB00002B/32